The SpringerBriefs in Space Development series explores the multifaceted field of space exploration and its impact on society. Under the editorship of Dr. Pelton and the auspices of the International Space University, the series features interdisciplinary contributions from space experts and rising professionals alike, offering unique perspectives on a rapidly expanding field.

The volumes in this series are compact, ranging from 50 to 125 pages (25,000–45,000 words). Instructors, students, and professionals will find herein a host of helpful reads, including snapshot reviews of a hot or emerging field; introductions to core concepts; extended research reports providing more detail than is possible in a conventional journal article; manuals describing an experimental technique or technology; and essays on new ideas and their impact on science and society.

The Briefs are concise, forward-looking studies on the space industry, broaching topics such as:

- Space technology design and optimization
- Astrodynamics, spaceflight dynamics, and astronautics
- Resource management and mission planning
- Human factors and life support systems
- Space-related law, politics, economics, and culture

Through the SpringerBriefs in Space Development, readers will learn of the amazing progress and key issues born from the international effort to explore space. The Briefs are published as part of Springer's eBook collections, and in addition are available for individual print and electronic purchase. They are characterized by fast, global electronic dissemination, straightforward publishing agreements, easy-to-use manuscript preparation and formatting guidelines, and expedited production schedules.

More information about this series at http://www.springer.com/series/10058

The SpringerBriefs in Space Development series explores the multifaceted field of space exploration and its impact on society. Under the editorship of Dr. Pelton and the auspices of the International Space University, the series features interdisciplinary contributions from space experts and rising professionals alike, offering unique perspectives on a rapidly expanding field.

The volumes in this series are compact, ranging from 50 to 125 pages (25,000–45,000 words). Instructors, students, and professionals will find herein a host of helpful reads, including snapshot reviews of a hot or emerging field; introductions to core concepts; extended research reports providing more detail than is possible in a conventional journal article; manuals describing an experimental technique or technology; and essays on new ideas and their impact on science and society.

The Briefs are concise, forward-looking studies on the space industry, broaching topics such as:

- Space technology design and optimization
- Astrodynamics, spaceflight dynamics, and astronautics
- Resource management and mission planning
- Human factors and life support systems
- Space-related law, politics, economics, and culture

Through the SpringerBriefs in Space Development, readers will learn of the amazing progress and key issues born from the international effort to explore space. The Briefs are published as part of Springer's eBook collections, and in addition are available for individual print and electronic purchase. They are characterized by fast, global electronic dissemination, straightforward publishing agreements, easy-to-use manuscript preparation and formatting guidelines, and expedited production schedules.

More information about this series at http://www.springer.com/series/10058

SpringerBriefs in Space Development

Allyson Reneau

Moon First and Mars Second

A Practical Approach to Human Space
Exploration

Allyson Reneau
M.A. International Relations
Harvard University
Oklahoma City, OK, USA

ISSN 2191-8171   ISSN 2191-818X   (electronic)
SpringerBriefs in Space Development
ISBN 978-3-030-54229-0   ISBN 978-3-030-54230-6   (eBook)
https://doi.org/10.1007/978-3-030-54230-6

This Springer imprint is published by the registered company Springer Nature Switzerland AG
The registered company address is: Gewerbestrasse 11, 6330 Cham, Switzerland

ISU Central Campus. (Credit: ISU)

This Springer book is published in collaboration with the International Space University. At its central campus in Strasbourg, France, and at various locations around the world, the ISU provides graduate-level training to the future leaders of the global space community. The university offers a two-month Space Studies Program, a five-week Southern Hemisphere Program, a one-year Executive MBA and a one-year Master's program related to space science, space engineering, systems engineering, space policy and law, business and management, and space and society.

These programs give international graduate students and young space professionals the opportunity to learn while solving complex problems in an intercultural environment. Since its founding in 1987, the International Space University has graduated more than 3,000 students from 100 countries, creating an international network of professionals and leaders. ISU faculty and lecturers from around the world have published hundreds of books and articles on space exploration, applications, science, and development.

NASA is recognized worldwide and carries with it the awe-inspiring achievements of dedicated men and women of the United States of America who have worked for the agency and accomplished what many thought to be impossible. A few of you have received recognition, but the majority of you have not. Therefore, to all of you, both past and present I dedicate this work. As it has been said, NASA turns science fiction into science fact.

Through all of NASA's ups and downs, the vision and passion to explore our solar system and beyond still burns brightly in the hearts and minds of its people. You have inspired all of us to reach for our dreams, never give up, and recognize that "Failure is not an option." For this, we all thank you.

Foreword I

How do we achieve a human presence on Mars? This brief but fascinating book addresses this question and proposes a logical answer. Explore the Moon first and do it now! The Moon is a wonderful space laboratory to learn and develop the next steps forward in inhabiting the red planet and the solar system. The numerous challenges of human space exploration are presented, the importance of international cooperation is emphasized, and the fact that we are on our way to the moon is presented. This book must be on your reading list!

Dr. Giovanni Fazio
Harvard University Senior Physicist
Smithsonian Center for Astrophysics

Foreword II

In the closing days of the Cold War between the United States and the Soviet Union, the race to space and to the Moon was part of a global competition between the world's two space superpowers. As Washington Post writer Christian Davenport has put it, these contenders for space superiority "…. bushwhacked a frantic path to the lunar surface, landing nearly 20 spacecraft softly on the Moon between 1966 and 1976, including the six carrying NASA's Apollo astronauts."

But then the Cold War receded into the annals of history and attention in terms of space exploration seemed to turn to the planet where spacecraft like Spirit and Opportunity (now defunct) and the still-functioning Mars Exploration Rover Curiosity have captured headlines. Entrepreneur Elon Musk as the head of Space X has found a bully pulpit to talk of how he foresees a day when there will be a fully functional human society of a million people operating a colony on Mars. Kids around the world dream of going to Mars since astronauts have already walked the Moon's surface.

But today, the world of space has spun back four decades to put the Moon back at the epicenter of space aspirations by the space powers. However, this new and exciting race back to the lunar surface has a new cast of characters. China has been the first to land a spacecraft (i.e., Chang'e 4) on the far side of the Moon. India, Russia, Japan, Europe, and China have other missions planned to go the Moon. The United States with its Lunar Gateway is clamoring to get back into the new lunar excitement.

This book, by space policy expert Allyson Reneau, about why there should be serious attempts to go back to the Moon on a more permanent basis is thus especially timely. Her book, which is derived from her thesis at Harvard University, involves interviews with leading space experts from around the world and answers some key questions.

It explains why such lunar initiatives at this time makes sense. It helps us understand what we can learn from lunar activities that will help us take the next steps beyond. Her book helps us understand what the Moon can teach us, in the next few years, about space and "NewSpace" enterprise. It can help us make sense of these initiatives from a technological, operational, financial, and physiological perspective.

The title of "The Moon Now, Mars Later" tells us all. This book is a refreshing review of all the reasons why the 2020s should be about lunar exploration, applications, and services. Then and only then, perhaps in the 2030s, we might know enough about space habitats and living in space (close to home) to take the next steps. In short, there is plenty of time to use the 2030s to go Mars, explore the moons of Jupiter, and perhaps even to use space systems in new and creative ways to save our planet from cosmic threats.

This being said, the way forward may still contain a few twists and turns. There are some who still feel that a focus on Mars exploration is the best advancement for space exploration. The U.S. Congress and the Trump White House are not fully aligned. There could be appropriation bills that dictate changed priorities and the U.S. presidential election in November 2020 could also lead to modifications. Regardless of such fluctuations, the wisdom of learning more from exploring the Moon before heading to Mars remains.

Until then we might do well to read about the research and practical ideas of Allyson Reneau about why the Moon and lunar activities are where the United States and other space-faring nations might best concentrate their attention. Let's keep our eyes on the ball and see where Space 2.0 activities may lead us. Let's hope it is toward more support from NewSpace players and better and more international cooperation in space. We will make the most progress if we learn how to share the benefits of space with the entire human community in the fastest and wisest sense possible.

Joseph N. Pelton

Founder of the Arthur C. Clarke Foundation and

Former Chairman of the Board of Trustees and Dean of the International Space University

Preface

Consensus is building worldwide for returning humans to the surface of the Moon. The United States, China, India, Russia, Canada, Japan, and the European Space Agency are setting their sights on our closest neighbor in outer space. From the beginning of time, the Moon has helped define our life here on Earth, as this celestial body is critically connected to us in a variety of ways. The Moon offers us enormous value scientifically as well as the possibility of rich, raw resources that can benefit our life on planet Earth.

It can also be a way station for further exploration and development related to more distant bodies in the Solar System. What we learn on the Moon can help us cultivate the technology, life support systems, operational capabilities, new regulatory policies, and space law critical for the future. The knowledge learned on the Moon in the 2020s could ultimately prove crucial to engaging in space mining, material processing, creating cost-effective space power systems, and even to the settlement of Mars in future decades.

Clearly there is much interest, excitement, and change now in the air concerning the future Lunar exploration. More than a dozen countries are actively engaged in one aspect or another of new developments related to launcher technology, logistical support or scientific equipment, and physiological research. There are at least five different cislunar rocket systems now in development. Change is also being fueled by national elections, new commercial initiatives, and especially the effects of the worldwide COVID-19 pandemic. The spread of the novel virus is impacting the global economy and especially space systems and exploration. In this environment, the exploration of the Moon can be seen as an optional area of investment by governments—especially in a time of economic recession. Thus, the impact of this current economic downturn remains to be seen, and subsequent economic recovery will allow lunar exploration to continue on pace in the 2020s. Despite the current shifts now occurring, the information presented in this book remains highly relevant. The information and analysis that follows in this book provides vital background about the where, what, and how of lunar initiatives now underway. The future effort to put humans on the Moon in permanent ways remains a key global objective. This exploration and perhaps planned commercial exploitation of the resources on the Moon, as well as its scientific exploration, will remain a key goal of the 2020s even though those objectives could now be somewhat delayed a few years.

This book will investigate the rationale for international space agencies as well as private companies to return to the lunar surface through international, public-private partnerships and even entrepreneurial initiatives. Although the horizon goal is Mars (and beyond), there has been a lack of political will, shrinking space budgets, and underdeveloped technological and medical knowledge to allow this to happen. All of these factors point the dial on the compass of space exploration in the direction of going back to the Moon.

The Moon should serve as a proving ground for a crewed mission to Mars due to the life-threatening hazards of living and working in deep space. It will allow us a great deal of new information about generation of power in space, space communications and information, operational protocols, and safety practices. This exploration will create governing policy and legal practices associated with human habitats and living in space from a practical and regulatory perspective.

The advancement of ambitious human space exploration needs to be done in an orderly and safe manner to preserve and protect life and sustainability of human habitats in this dangerous and exciting quest.

> The space program has never been an entitlement, it's an investment in the future - an investment in technology, jobs, international respect and geo-political leadership, and perhaps most importantly in the inspiration and education of our youth. Those best and brightest minds at NASA and throughout the multitudes of private contractors, large and small, did not join the team to design windmills or redesign gas pedals, but to live their dreams of once again taking us where no man has gone before. (Gene Cernan, Apollo 17 Astronaut, September 2011)

M.A. International Relations   Allyson Reneau
Harvard University
Oklahoma City, OK, USA

Acknowledgments

I would like to thank my incredible mother and father, my 11 beautiful children, and my steadfast friends; the inspiring deans, advisors, and faculty at the University of Oklahoma and Harvard University; and my loving God. During difficult times, each of you whispered in my ear "You can do it!" For this I will always be grateful. It is to all of you I dedicate this work, as well as the many opportunities my education has afforded me. I promise to make this world a better place.

Contents

Chapter 1
Introduction to Space Exploration

In a historic speech to a joint session of Congress on May 25, 1961, President John F. Kennedy boldly announced his ambitious goal of putting Americans on the Moon by the end of the 1960s. A little more than 8 years later, the Apollo 11 mission would successfully land the first two humans on the Moon. Sadly, President Kennedy would never see the fruition of this daring and dramatic Moon-shot speech. He was assassinated on November 22, 1963, while driving in a presidential motorcade in downtown Dallas, Texas. And just as America's young and promising president died young, so in time, NASA's aspirations for an active human space program beyond the confines of Earth's orbit, withered away in the decades to follow.

What caused the premature truncation of the US deep space exploration goals in the early 1970s after one of the greatest achievements of our human existence? This book will explore the conception and birth of the US human deep space program, its accelerated and flourishing first decade, followed by the spurts and sputters over the last 50 years that have followed. The current presidential administration has tasked NASA with a transition of the government-funded International Space Station to a commercially viable human presence in low Earth orbit. In addition, as a result of activities of the reconstituted National Space Council, NASA was recently commissioned with the new and challenging schedule of returning humans to the lunar surface by 2024. The question on the minds of veteran professionals in the space industry is the following: Will this outcome be different from the pro-Moon initiatives of the father-son duo of President George H.W. Bush and President George W. Bush that ultimately fizzled out? These programs to take human crews back to the Moon and on to Mars both failed due to underfunding, lack of congressional support, schedule overruns, and the start-stop cycle of the change of presidential leadership in the United States.

The question now is whether there is a global race to the Moon as was the case in the 1960s with competitive space programs from the United States and the Soviet Union. The answer is yes. There are competitive space initiatives involving sending

© The Editor(s) (if applicable) and The Author(s), under exclusive license to
Springer Nature Switzerland AG 2021
A. Reneau, *Moon First and Mars Second*, SpringerBriefs in Space
Development, https://doi.org/10.1007/978-3-030-54230-6_1

exploratory and human-crewed missions to the Moon, but it is a totally different competition—though in some ways, just as fierce.

Space historian and space policy expert, Dr. John Logsdon, often proposes the question of "Will we have another 'Kennedy moment' that awakens the US from its seemingly laissez-faire performance--when it comes to human space exploration?" The answer at this moment in the twenty-first century might be a resounding "Yes!" In fact, the Moon race has already begun with China and Russia challenging the United States for global supremacy in many areas. The current philosophy of the American space program is once again to "Think big." The goal outlined in the US Space Policy Directive One is a call for "America to lead in space again." This aspiration for the United States to lead in space has caused a major shift and course-correction in current American space policy. Today there are not only a number of national initiatives from the United States, China, Europe, India, Israel, and Russia to explore the Moon, but there are a growing number of private enterprise ventures as well.

The private "NewSpace" initiatives are also progressing at an extraordinary rate. These commercial space companies led by billionaires, ambitious innovators, and space entrepreneurs are discovering new ways to travel to space with a much smaller price tag—and often do so much faster and more efficiently. Merrill Lynch predicts that the 2020 space economy will top out at nearly 400 billion, with current projections of space-related businesses perhaps progressing to three trillion by 2050 (Sheetz Oct. 31, 2017).

There are other estimates that are built on an analysis of various new space services and space-related activities that suggest that the increase from 400 billion dollars to a trillion-dollar economy will come within perhaps a trillion dollars in a little over a decade (Pelton 2019).

And now, to save time and tax-payers money, NASA is becoming a customer to these companies. For instance, Boeing and SpaceX have begun launching NASA astronauts to the International Space Station. Space entrepreneurs are already planning human orbits around the Moon, lunar habitations for research and tourism, as well as a private journey to Mars. The world is witnessing a unique and exciting renaissance. These strictly private space initiatives are addressed in later chapters.

The premise of this book is that it is important to lay a strong and safe foundation of solid technology, space power systems, viable telecommunications, information technology, sustainable infrastructure, and proven life-support systems at the lunar surface before we humans seek to venture to Mars and beyond. The Moon can serve as a proving ground for longer termed crewed missions to Mars.

There are many rational justifications as to why the Lunar Gateway and lunar habitats are the best locale for proving that humans can sustainably survive the life-threatening hazards of living and working in deep space. There are numerous good reasons why a test lab that is only 238,000 miles/400,000 kilometers from Earth is better than one that is many millions of miles/kilometers away. As space-faring nations and commercial players set their sights on a path of ambitious human exploration, they should do this in a well-ordered and progressive manner. This is not only to preserve this visionary ideal but, most importantly, to protect human life.

Despite the many reasons given above for learning about how to sustain life on the Moon via NASA's Artemis program, plus the many other initiatives now planned by Europe, India, Japan, China and Russia as presented later in this book, it is important to note that priorities may still shift. There are those that believe prioritizing Mars missions still make sense. This is to say that some aspects of these efforts could still be changed to some degree, depending on the outcome of elections in the United States in 2020. This would mean that perhaps some aspect of these lunar programs might be altered. It is, however, considered much more likely that the lunar exploration programs described in this book, at least in general terms, represent the basic outline of the US and other space nations' human space initiatives to be expected in the next decade.

A Personal and Unique Perspective

Prior to being directly involved in the world of space, technology, and national space policy, my life was what I would call "normal." As a typical middle-class mother living in the heartland of the United States, NASA and US space policy were not on my mind, and not a priority for me or the other people who resided in my community. As a mother of 11 children, my daily focus was *not* on going to the Moon or Mars. It was getting the kids to school on time, feeding hungry mouths, paying the mortgage, and taking my crew to extracurricular sports and music events—all while running a full-time business. No one in my neighborhood or city thought or talked much about space exploration or technology. My taxpaying friends and local citizens (if they thought about space at all) wondered why the government should spend enormous amounts of money on space exploration when there are so many problems on Earth.

After studying and researching space policy at Harvard University, spending time at NASA Headquarters in Washington DC, and also receiving a degree from International Space University, I have had a special opportunity to view the issues related to human and robotic space exploration, space sciences, and space applications (satellites). It is clear to me now the amazing benefits of the ambitious pursuit of "space." All of the money spent on space programs generate jobs and salaries paid directly to people working here on Earth. The knowledge that comes from space programs develops beneficial expertise that makes our everyday life easier and more comfortable. The Merrill Lynch projection of a three trillion-dollar space economy by 2050 will translate into jobs created, money earned, and a bolstering of new economic activities—all benefiting the global economy. I hope to address these questions, because I am sure you have had the same thoughts in your mind.

Perhaps after reading this book, you will see that continued investment in space exploration yields tangible and, more importantly, intangible rewards. In particular, human space exploration holds the promise of generating a return on investment for us in terms of new technology and knowledge creation. This, in turn, can provide solutions to unanswered problems that citizens around the world face on a daily basis. Some of these dilemmas which space systems and exploration help solve may include cures for deadly and disabling diseases; better communication; cleaner energy; better warning systems for natural disasters including tornadoes, earthquakes, and hurricanes; synchronization of the internet; location of vital minerals to mine; and planetary protection against solar storms and asteroid strikes. Every time one takes off or lands safely on a plane, utilizes a navigation system to get to a destination in a car, watches their favorite television show, or uses their smart phone, they are probably benefitting from the application of space systems at work far above the earth.

The technology that has been developed through space programs has been transferred to our everyday lives and is at work in so many places we don't even realize. Space-based innovations can be seen almost everywhere one looks from the operating room, to "smart" agriculture, to the design and construction of homes, offices, hotels, and skyscrapers which are built with stronger more resilient building materials. The space-based "spinoffs" are now so prevalent that most people (including me a few years ago) are completely unaware of the positive aspects of space exploration that affect many hours of their normal day. To better grasp these extraordinary benefits, one could go to YouTube and a find a short video that shows by example all of the vital services that people around the world would lose if we were suddenly denied access to our communications, navigational remote sensing, weather, and defense system satellites. The title of this video is *"If There Were a Day Without Satellites"* (If There Were…2016).

The purpose of this book is to consider the fact that our Moon is a wonderful space laboratory. It is close at hand. It is the easiest neighbor to reach and to communicate with a minimum delay and at the lowest cost. If humans are to execute activities in space for the long term—and ultimately utilize the resources of space—it is logical and economically smart to begin with the Moon first. This book thus seeks to define and evaluate the best pathways forward for human exploration and infrastructure development in outer space.

The analysis, opinions, and knowledge of many space experts will help answer important questions about the next steps forward in space exploration, applications, and development. Their shared expertise will assist in explaining what achievements are possible in the coming years and decades, including when and what to do first and why many of the key progressive steps can and will take place on the Moon. This exploration of our closest neighbor will reveal the next logical steps forward that are important for space agencies around the world to appraise, evaluate, and act on in the decades ahead. It is also equally important, for those in the commercial space industries to consider the opportunities and act accordingly.

In 1958 Martin Luther King Junior said, "Human progress is neither automatic nor inevitable. Every step requires sacrifice, suffering, struggle, and the tireless exertion and passionate concern of dedicated individuals" (Martin Luther King Jr. 1958).

This is true of most worthwhile ventures, and it is true of space initiatives of substance. Investment in space coupled with tireless and passionate exertion can now pay off generously in the future. If the three-trillion economy in space is achieved in 2050, it will all become clear why such investments made sense.

References

If There Were A Day Without Satellites. (2016). *YouTube*. https://www.youtube.com/watch?v=5sgM7YC8Zv4. Last accessed 2 Feb 2020.

Martin Luther King Junior. (1958). https://whatwillmatter.com/2017/01/worth-reading-michael-josephsons-martin-luther-king-day-speech. Last accessed 3 Feb 2020.

Pelton, J. (2019). *Space 2.0: Revolutionary advances in the space industry* (p. 2). Cham: Springer Praxis Press.

Sheetz, M. (Oct. 31 2017). *The space industry will be worth nearly 3 trillion in thirty years Bank of America predicts*, CNBC.com. https://www.cnbc.com/2017/10/31/the-space-industry-will-be-worth-nearly-3-trillion-in-30-years-bank-of-america-predicts.html. Last accessed 15 Jan 2020.

Chapter 2
Background and Early History of Space Exploration

On July 20, 1969, astronauts Neil Armstrong and Buzz Aldrin stepped on the Moon, and with those steps, they accomplished a great historical triumph. Subsequent missions to explore the lunar surface continued, with Apollo 17 astronauts Gene Cernan and Harrison Schmidt leaving the last human footprints on the Moon in December of 1972. Since that time, there have been many Space Shuttle missions and an extensive human presence in space on the International Space Station (ISS). These are, no doubt, significant accomplishments in space and have created positive results for all on Earth. But, for the past 40 years, space travelers have been trapped in the repetitive cycle of low Earth orbit (LEO) space missions. If humans are to explore this solar system and universe, there must be an exit from this pattern. Ultimately, there is a need to escape the gravity well of Earth and once again reach for the stars—or at least Mars and beyond—if humanity is to achieve its ultimate destiny (Chaikin 1994).

Human exploration has inspired mankind since the beginning of civilization. Exploration is one of the characteristics of a progressing society. When a nation explores into the unknown, it leads to innovation, discovery, and, ultimately, prosperity. In addition, it can also foster national prestige, inspire its citizens to pursue excellence, and create new knowledge and initiatives for the future. What is the next logical step for spacefaring nations? Is it a return to the Moon, or a direct trip to Mars—or some yet to be identified next step forward?

What follows is a step-by-step evaluation of the advantages and disadvantages of each of these choices. Let's analyze this question by looking deeply into six evaluative categories central to the future of space exploration and discovery: physiological, psychological, technological, economical, international, and national. There is a good reason to believe that these six areas are possibly the most critical for making a suitable decision. Arguably, there are other important categories that exist, but a consideration of these critical subgroups should render what may be a qualified and sensible path forward. Intermingled with these six categories is the legal and regula-

© The Editor(s) (if applicable) and The Author(s), under exclusive license to
Springer Nature Switzerland AG 2021
A. Reneau, *Moon First and Mars Second*, SpringerBriefs in Space
Development, https://doi.org/10.1007/978-3-030-54230-6_2

tory environment that allows progress in these areas to be achieved. Thus, relevant legal and regulatory issues will be addressed throughout the book as this analysis continues.

The rich space exploration history over the last several decades is relevant here as nations around the world consider what to do next. This is key in that humanity seems to be at an important crossroads of decisions concerning the long-term sustainability and quality of life on Earth. The future of space and key issues such as sustainability of life on Earth are, in fact, deeply intertwined. The continued investment in human and robotic space exploration spawns new scientific and engineering knowledge. It also generates priceless symbolism and extraordinary educational and research inspiration from which people and nations can create meaning and purpose. If this is done for the right reasons and in a spirit of international cooperation, it can possibly develop a more united identity for humanity. Space exploration could ultimately serve to reduce the human tendency to compete through acts of war and violence.

This inspirational function can be particularly true for professional women. The following quote comes from Kathy Laurini, NASA, Senior Advisor, Exploration and Space Operations. She is just one of the tens of thousands that have been inspired by the intellectual, educational, and philosophical inspiration from being involved in what might be call the "space adventure."

> I was nine years old during the landing of Apollo 11, when astronauts Neil Armstrong and Buzz Aldrin made the first steps on the Moon, and it influenced me to be an engineer. It stopped the world! I have given NASA tours to my friends, and then their children went on to become engineers and scientists. Space inspires the lives of people, and we need to communicate everything we accomplish, because creating this awareness pays large dividends in adults, as well as our children. (Laurini July 28, 2015)

Space in the Early Beginnings of the Cold War

After the conclusion of World War II in 1945, a new ideological battle, termed the Cold War, began to emerge between the world's two superpowers—the communist Soviet Union and the democratic United States. Each country attempted to show superiority, both militarily and technologically, hoping to prove that its socialistic or capitalist system, respectively, was the best. As the rivals contended for supremacy, outer space became a spectacular arena for their competition (John F. Kennedy Presidential Library and Museum February 21, 2016).

On October 4, 1957, the Soviet Union struck a defining blow to the US outer space aspirations with the launch of Sputnik 1, the first artificial satellite, into space. This satellite launch not only inaugurated the Space Age but also the start of the "Space Race." This space competition was fueled by the Cold War tensions between the United States and Russia to a red-hot level. This leading and dominant technological achievement as represented by the Sputnik launch surprised and embarrassed the Americans, proving that the Soviets were, at the time, more advanced technologically in space systems. This unexpectedly early Sputnik launch and those that followed exposed to the political and military leadership the extent to which the United States was behind in the space race (Sputnik, NASA History 1957) (see Fig. 2.1).

A number of hurried attempts to launch a US satellite into orbit followed on for a few months. But these attempts, rather embarrassingly, ended in launch failure. Almost 4 months later in a hasty sprint, the United States launched its first satellite named Explorer on January 31, 1958. This mission had an on-board Geiger counter which detected the Van Allen Belts and could be considered first experimental scientific mission. The picture below shows from left to right, Dr. William Pickering, head of the Jet Propulsion Labs that built and operated the satellite; Professor James Van Allen of Iowa State University who the created the Geiger counter payload; and Werner Von Braun, who designed and oversaw the Jupiter C launch vehicle that launched Explorer I in 1958 (see Fig. 2.2).

Thereafter, however, the Soviets began to accomplish a series of ideological and technologically charged "firsts" from 1957 to 1961. These firsts included launching a dog into orbit on Sputnik 2. This was followed by perhaps the most spectacular launch achievement of that time by putting Russian Cosmonaut Yuri Gagarin into an Earth orbit. Although the first man launched into outer space and orbited the world only once before de-orbiting on April 12, 1961, the world press coverage was enormous (see Fig. 2.3).

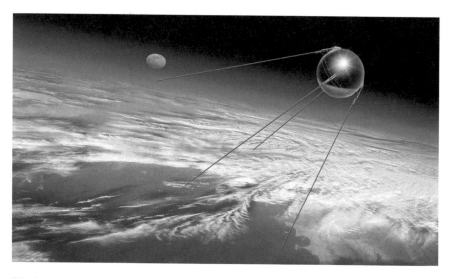

Fig. 2.1 Sputnik 1 above the Earth. (Graphic courtesy of NASA)

Fig. 2.2 The three architects of Explorer 1—the first successful US satellite launch. (Graphic courtesy of NASA)

Additional leading-edge accomplishments by the Soviet Union included the first woman in space, the first spacewalk, the first spacecraft to orbit the Moon, and the first unmanned spacecraft to land on the lunar surface. Propelled by these Soviet exploits, the US government and the military, science, medical, and technological communities united their efforts to move forward quickly in a space race to response. They feared the Russians might be considering more belligerent plans involving the creation and use of this new engineering advancement in space (Hanes 2012).

Fig. 2.3 Photo of Cosmonaut Yuri Gagarin— the first person launched into space. (Graphic courtesy of NASA)

References

Chaikin, A. (1994). *A man on the moon*. New York: Penguin Books.

Hanes, E. (2012). *From sputnik to spacewalking: Soviet space firsts. History in the headlines*. http://www.history.com/news/from-sputnik-to-spacewalking7-soviet-space-firsts. Last accessed 23 Nov 2019.

John F. Kennedy. Presidential Library and Museum, "The Cold War." http://www.jfklibrary.org/JFK/ JFK-in-history/The-Cold-War.aspx. Last accessed 21 February 2016.

Laurini, K. (July 28 2015). NASA, Senior Advisor. *Exploration and space operations*. Personal interview. International Space University.

Chapter 3
Maturity of the US Space Program

The Creation of NASA

The United States was both fearful and humiliated by the Soviet successful launch of Sputnik 1 in October 1957. America had, since World War II, prided itself on being superior to all other nations, militarily, scientifically, and technologically. This sense of urgency to regain a lead in space supplied President Dwight D. Eisenhower with sufficient political will and US Congressional support to create the National Aeronautics and Space Administration (NASA) on July 29, 1958 (Dick 2008).

Although this new civil space organization was designated as a civilian entity with the purpose of peaceful pursuits of outer space, President Eisenhower and his successor, President Kennedy, created additional national security space organizations that would operate in conjunction with NASA. The first series of actions were the undertaking of intelligence satellites Corona, Argon, and Lanyard, programs operated serially from the late 1950s to the early 1970s. These high-resolution space surveillance programs produced strategic reconnaissance satellites to obtain intelligence on the Soviet Union (Skytland 2012).

These satellites collected information on bombers as well as on space and missile capabilities. The second-related action was the creation of a new secret entity known as the National Reconnaissance Office, (NRO), established in 1961, which was committed to utilizing outer space for the potential military uses and especially detailed scrutiny of strategic information through the analysis of reconnaissance satellite information.

© The Editor(s) (if applicable) and The Author(s), under exclusive license to
Springer Nature Switzerland AG 2021
A. Reneau, *Moon First and Mars Second*, SpringerBriefs in Space
Development, https://doi.org/10.1007/978-3-030-54230-6_3

Project Mercury

One of the first goals of NASA was to launch a man into Earth's orbit as soon as possible. The initial objective of this space program was to study a human's ability to perform in space and return safely from orbit. The program to accomplish this task was named Project Mercury and lasted from 1958 to 1963 (see Fig. 3.1).

Seven astronauts were selected from the military's test pilot program, and these men were known as Mercury Seven. On May 5, 1961, Astronaut Alan Shepard was launched into a 15-minute suborbital space flight aboard Freedom 7. In early 1962, John Glenn became the first American to be launched into Earth's orbit. As already noted, this was a feat accomplished by the Russians nearly a year earlier (Dunbar 2015a).

All said and done, Project Mercury launched 20 unmanned flights (although several included animals), and 6 flights were launched that carried astronauts. The strategy and goals of Project Mercury were very distinct—and successful—even though the Soviet Union launch operations managed to put a man into Earth's orbit before the United States.

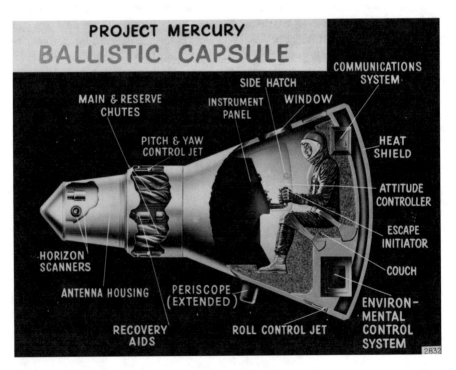

Fig. 3.1 The Project Mercury Capsule. (Graphic courtesy of NASA)

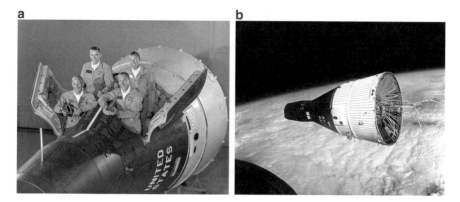

Fig. 3.2 (**a, b**) Gemini Astronauts and the Gemini Capsule. (Graphics courtesy of NASA)

Project Gemini (Fig. 3.2)

Project Gemini was a human spaceflight program conducted by NASA in 1965 and 1966, another prelude to the upcoming Apollo program. This program was an upgrade from Mercury Program with a larger capsule that held a crew of two astronauts. Not only did the Gemini Program include all of the Mercury Seven astronauts but also a second class of astronauts named "The New Nine" and then a third astronaut class chosen in 1963. The goals of Gemini were very specific, and the main objectives were to master and prove several space travel maneuvers to support a manned mission to the lunar surface. These maneuvers included astronauts working in space outside the spacecraft (EVA's or Extra Vehicular Activity), living and working in space for the same number of days it would take to go to the Moon and back, and orbital rendezvous—which involved docking two spacecraft together in space. In addition, reentry techniques were perfected, and learning was expanded with regard to specific touchdowns at specific locations. These new abilities accomplished by Gemini gave the Apollo mission the necessary techniques to pursue its goal of human exploration to the Moon (Hitt 2011).

The Mercury and Gemini Programs also moved to create expanded tracking, telemetry, and command capability (TT&C) to use not only ground-based tracking stations but also to use satellites and ship-mounted tracking systems. The Intelsat II satellite program was commissioned by the US government to augment such space-based tracking capabilities over the Atlantic Ocean.

Project Apollo

The US citizens' perception of Soviet superiority in space systems was one of the prime incentives needed for newly elected President John F. Kennedy to make a special request to Congress. This young president proposed that by the end of the

1960s, the US government would commit to putting a man on the Moon and returning him safely home. This speech was bold, challenging, and a strategic move on the part of President Kennedy, and it was not immediately embraced with enthusiastic national and political support. Nevertheless, the social and political foundations were laid for a manned lunar landing, which led to the establishment of the ambitious Apollo program. Although the United States was behind in the space race, Americans found themselves suddenly engaged in a dramatic competition in which they were determined to be the victor and the first to send astronauts to the Moon.

NASA's budget was immediately increased. The NASA budget for Apollo rapidly grew as the urgent Apollo mission demanded more and more funds. The NASA budget quickly swelled by nearly 500 percent, and the Mercury, Gemini, and Apollo program soon became the most expensive scientific endeavor undertaken by the United States. NASA's expenditures in the 1960s were nearly $25.4 billion (about $250 billion in 2020 economic terms). Apollo employed more than 400,000 people from both within NASA, as well as from a large army of civilian contractors to pursue this daunting goal—and to do it on schedule (NASA, Project Apollo 2014).

The schedule was almost dizzying in its urgent demands. During the period from 1961 to 1963, NASA completed the Mercury Program. Next came the demanding Gemini program which lasted from 1965 to 1966. The last part of this tripartite civil space program to take astronauts to the Moon lasted through December 1972. It was then decided by the Nixon administration to cancel the Apollo 18, 19, and 20 missions, which grounded the final Saturn V rockets, and the entire Apollo program.

Meanwhile, the secret Soviet moon landing program (declassified in 1990) was losing political momentum, as the Apollo program was ramping to its highest levels. There were internal tensions within the Russian government space organization due to administrative, technical, and financial difficulties and was further delayed by the unexpected death of the chief space engineer, Sergey Korolov, in 1966. The Soviets attempted four launches designed to achieve a Moon landing during the period from 1969 to 1972. These missions were not widely publicized and failed each time. These factors when taken together, led eventually to the cancellation of the Soviet lunar program (David February 7, 2011).

The Apollo program was not all smooth sailing either. The electric spark that occurred in the Apollo Command Module during a test with three astronauts on board created a disastrous and deadly fire in the pure oxygen atmosphere, with highly combustible paper acting as the igniting fuel. But the program was restructured, and safety engineering came to the forefront. After a series of uncrewed flights, Apollo 7 was launched in October 1968 with astronauts on board. Following the further successful flights of Apollo 8, 9, and 10, the goal of astronauts landing on the Moon before the decade's end finally seemed to be achievable. The Apollo Command and Service Module is shown in Fig. 3.3.

On July 16, 1969, Apollo 11 launched on a Saturn V rocket from Cape Canaveral, Florida. The launch was the first attempt of a manned lunar landing. American astronauts Neil Armstrong, Edwin "Buzz" Aldrin, and Michael Collins were aboard the spacecraft heading for the Moon. When Neil Armstrong set foot on the Moon for the first time in human history, he spoke these famous words, "One small step for

Fig. 3.3 The Apollo Command and Service Modules shown above the Moon. (Graphic courtesy of NASA)

Fig. 3.4 Image from the 1969 Moon landing. (Graphic courtesy of NASA)

man, one giant leap for mankind." From the start, American citizens were captivated by this journey to the Moon, and television caused interest in the phenomenon to spread around the planet, with nearly 600 million viewers watching the landing (Dunbar 2015a) (see Fig. 3.4).

This large global audience was only possible due to commercial satellite communications. Just 10 days before the landing, the Intelsat global satellite network moved an Intelsat III from the Pacific to the Indian Oceans region so that the television of the Moon landing could be seen around the world. The signal from the Moon was first received by the Parkes radio telescope in Australia and then relayed to various Intelsat satellite earth stations that could connect to all regions of the world. The television signal was sent to Mission Control in Houston and forwarded to television networks worldwide. Ultimately, the satellites provided global connectivity to create the largest television audience in human history up to that time (Pelton et al. 2004).

Because of these groundbreaking satellite systems, the Apollo 11 astronauts became worldwide heroes. Although they placed an American flag on the Moon, they ostensibly came on behalf of all mankind. The global satellite distribution of this event on television seemed to make this a globally shared experience. Adults and children alike were inspired all over the world, and the Americans shared a strong sense of pride in this extraordinary accomplishment.

To many citizens in the United States and abroad, the NASA landing of a man on the Moon tended to indicate that the United States was the winner in the space race against the Soviet Union. The US government's passion for further human space exploration soon began to wane. But one last major space accomplishment remained.

In 1975, a joint US and Soviet space mission, named Apollo-Soyuz, was successfully staged in low Earth orbit. For this project, the United States sent three American astronauts to space in a surplus Apollo capsule, and the USSR sent two Soviet cosmonauts into space in a Soviet spacecraft. The two spacecraft docked successfully on July 17, 1975. The commanders of the mission, Astronaut Thomas Stafford and Cosmonaut Alexey Leonov, greeted each other with an official handshake in space (Ezell 1978).

This was followed by a signing ceremony of the joint mission certificate between the United States and the USSR (see Fig. 3.5).

This joint mission became one of the several symbolic events that signified the approaching end of the current Cold War, as the United States and Russian political relations began to improve—although it can be argued that the Cold War did not officially end until the USSR collapsed some 16 years later.

As the space race thawed—along with the easing Cold War—the United States' interest in space exploration continued, but at a much slower pace. The public became less excited with five repetitious missions to the Moon, and NASA was losing momentum. NASA needed a new goal.

President Richard Nixon was not motivated to set another ambitious, risky, and costly plan like Apollo. Nixon viewed NASA as another domestic government agency competing for taxpayer dollars and no longer a "favored" program. The lasting impact of Nixon's space doctrine was to terminate human space exploration which thereafter locked humans into low Earth orbit for decades to come (Logsdon 2008).

Space Shuttle Program (Fig. 3.6)

In 1970, after much thought and counsel, President Nixon decided to cancel the three remaining lunar missions, Apollo 18, 19, and 20. Instead, he favored shifting NASA's decreased budget to a new sustainable program: the creation of a partially

Fig. 3.5 The United States and USSR signing the Joint Mission Document in 1975 for the Apollo-Soyuz programs. (Graphic courtesy of NASA)

Fig. 3.6 The Space Transportation System or Space Shuttle. (Graphic Courtesy of NASA)

reusable spacecraft named the Space Shuttle, formally termed the Space Transportation System (STS) program. The Space Shuttle was capable of reaching LEO, and it flew 135 missions between 1981 and 2011. Two shuttles, Challenger in 1986 and Columbia in 2003, were lost in tragic accidents. Altogether, 14 astronauts perished in these 2 events, plus 3 more had previously perished in the Apollo 1 fire. In contrast, only four cosmonauts were lost in their crewed space missions (Heppenheimer 2004).

During this 20-year period, the Space Shuttle launched several major satellite missions, the Hubble Space Telescope (HTS), as well as the Hubble repair missions. Beginning in 1988, the Space Shuttle initiatives also led to other means of international cooperation in space. One of the most important of these international activities was the 1983 launch of Spacelab, an orbiting reusable laboratory in space, housed in the cargo bay of the Space Shuttle. In the Spacelab, research and scientific experiments were conducted in several areas including microgravity, human performance in space, astronomy, life sciences, orbital sciences, etc. The Space Shuttle program facilitated and, in many ways, set a new precedent for international relationships in space within a multidisciplinary setting. Altogether, 22 Spacelab missions were flown aboard the Space Shuttle from 1983 to 1988. The Shuttle also launched the majority of the components for the International Space Station.

Space Stations

In 1973, NASA launched a space station named Skylab, which was a science laboratory and solar observatory. Three manned missions transporting three astronauts were conducted on Skylab, and hundreds of physical and life science experiments were conducted while in low Earth orbit (see Fig. 3.7). Skylab was eventually guided into Earth's atmosphere in 1979 and disintegrated (Armstrong 2003).

After Skylab, NASA created more space laboratories, including Spacelab and Shuttle-Mir, a cooperative space venture with Russia's Mir space station. Then the concept of Space Station Freedom was considered under the presidency of Ronald Reagan, who was in office from 1981 to 1989. Space Station Freedom, however, was never built as originally envisioned but would eventually evolve to become a component of the International Space Station and would inspire President Reagan to invite US allies and friends to participate in the program—which was a defining feature of the ISS (President Ronald Reagan January 25, 1984) (see Fig. 3.8).

Construction of the International Space Station (ISS) began during President Bill Clintons' second term as president late in 1998. The ISS became the replacement project for Space Station Freedom, and it would be deployed to a different orbit, in order to facilitate Russian participation, and in partnership with many international space agencies. The ISS formally became the official undertaking of five space agencies that included NASA (United States), ESA (Europe), JAXA (Japan), Roscosmos (Russia), and CSA (Canada) and is anchored by an Intergovernmental Agreement (IGA). The first element of the ISS was launched November 20, 1998, and has since included the active participation of 18 nations and has been visited by 240 plus astronauts and cosmonauts from 20 different countries. The International Space Station is a crewed platform the size of a football field and weighs nearly 460 tons. It took 30 flight missions and over 10 years to complete, with the first

Fig. 3.7 Skylab pictured above Earth. (Graphic courtesy of NASA)

Fig. 3.8 The Fully Deployed International Space Station as constructed by the United States and its many international partners. (Graphic courtesy of NASA)

American-Russian crew living several months on the station beginning November 2, 2000. The International Space Station has now been operational with a crew aboard for over two consecutive decades.

The ISS currently serves as a space and microgravity research laboratory where experiments are conducted by astronaut crew members. The experiments have been devised by an international assortment of research scientists, academics, and students drawn from around the world. Experiments have been conducted in the fields of biology, botany, chemistry, human physiology, medicine, meteorology, astronomy, as well as many others. It also serves as an important microgravity experimental human laboratory, proving ground, and simulator for future trips to the Moon, Mars, and other deep space destinations.

Constellation Program

Following the Columbia Space Shuttle disaster in 2003, President George W. Bush, on January 14, 2004, sought to regain public trust and enthusiasm for manned space flight. By 2002, NASA Administrator Michael D. Griffin helped create a human space exploration roadmap known as the "Exploration Systems Architecture Study," which was formalized into law by the NASA Authorization Act of 2005 and named the Constellation program (Connolly 2006).

Constellation was NASA's plan to return humans to exploring in space beyond low Earth orbit. The program began with the idea of exploring Mars and identifying what would be needed to accomplish this goal safely and effectively. Constellation called for three design reference missions: (1) the Orion crew capsule would dock with the ISS and stay in orbit for 180 days or more and then return the crew back to Earth; (2) the mission called for lunar sorties or short 7-day missions to the Moon; and (3) the missions would have crews staying on the Moon for 6 months at a time and establishing a lunar outpost (see Fig. 3.9).

In past generations, going to the Moon was the extent of US exploration, but the Constellation program included the possibility that the launch and crew carrying systems could also be employed for a future journey to Mars (Connolly November 4, 2014).

President Obama Space Policy

When President Barack Obama took office in 2009, he called for an in-depth review of current US space policy. To undertake this review, he appointed a high-level commission headed by former president and CEO of Lockheed Martin Corporation, Norm Augustine, and many other well-known space experts. The report was known as the Augustine Commission Report—or more formally, "Seeking a Human Space Program Worthy of a Great Nation" (Bonilla 2009) (Fig. 3.10).

The commission's findings included three major conclusions: (1) They proposed that commercial space industry and "NewSpace" entrepreneurial companies undertake and do much more of the new development and innovation for low Earth orbit (LEO) activities. It suggested that NASA convert much more of this type of near-Earth activity to industry, including transporting astronauts to and from the

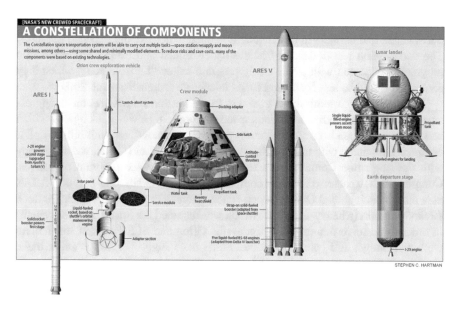

Fig. 3.9 The many component parts of the Constellation program. (Graphics courtesy of NASA)

Fig. 3.10 President Obama at Kennedy Space Center. (Graphic courtesy of NASA)

International Space Station. (2) They concluded that the current program of record, Constellation, had some key flaws. Thus, they proposed that the Constellation Ares launch vehicles and Orion spacecraft design and manufacture should be halted. These spacecraft designed to go to the Moon and ultimately to Mars were behind schedule, over budget, and unlikely to be adequately funded to achieve the intended goals. (3) The commission recommended that the under-funded and troubled Constellation program be replaced with a less costly program called "Flexible Path" which they contended would be less costly and easier to achieve (Augustine et al. 2009).

The Augustine Commission's goal was to ensure that NASA was on an aggressive yet sustainable path of ambitious space exploration. However, Congress, which had already endorsed the Constellation program and now had multibillion-dollar contracts associated with it, including Ares V (now SLS) and Orion, was not happy with the Augustine Commission's redesigning of the space program objectives. Congress in 2010 elected to retain components of Constellation as part of NASA's new plan. This included Orion (a deep-space, manned exploration vehicle) and the Ares V program (a heavy-lift rocket which was renamed the Space Launch System) in order to preserve this program. Today the Orion and SLS Program still remain well over budget and behind schedule. Some have suggested that it might make more sense to use private rockets and spacecraft to go to the Moon rather than continuing to fund the SLS and Orion programs. At this late stage, this seems unlikely.

President Obama was left torn by the conflicting advice. He proposed that NASA send astronauts to a near-Earth asteroid by 2025 and explore Mars by the mid-2030s. In addition, he extended funding for the ISS through 2024 (President Barack Obama April 15, 2010).

Not unsurprisingly, the aerospace community and program managers at NASA were frustrated by what they saw as the lack of credible advice by the Augustine Commission and a major compromise by President Obama. John Connolly, NASA's Chief Exploration Scientist, reacted to the Obama administration's political decisions by expressing his concerns:

> NASA wondered what they would do with these parts and pieces they had been given. What in the world do we do with these random pieces that don't really fit together? There is no real vision. It's just pieces and parts. I am a 'destination guy.' And now I have no real drive or motivation to move forward. The lesson that I have learned is that no matter how much money or energy we at NASA put into a project—the White House can shut it down. (Connolly November 4, 2014)

Similarly, Apollo 11 Astronaut Neil Armstrong commented, "When President Obama released his budget for NASA, he proposed a slight increase in total funding......and the accompanying decision to cancel the Constellation program, Ares I and Ares V rockets, and the Orion spacecraft is devastating." (Neil Armstrong April 14, 2010)

Likewise, Apollo 17 astronaut Gene Cernan remarked on the 2010 Obama space policy:

> For the United States, the leading space-faring nation for nearly half a century, to be without carriage to low Earth orbit and with no human exploration capability to go beyond Earth orbit for an indeterminate time in the future, destined our nation to become one of second or even third-rate stature. While the President's plan envisages humans traveling away from Earth and perhaps towards Mars sometime in the future, the lack of developed rockets and spacecraft will assure that ability will not be available for many years. (Cernan April 15, 2010)

Robert Zubrin, president of the Mars Society, also criticized Obama's plan in an article in the 2010 New York Daily News:

Under the Obama plan, NASA will spend $100 billion on human spaceflight over the next ten years in order to accomplish nothing. The President called for sending a crew to a near-earth asteroid by 2025. Had Obama not cancelled the Ares V, we could have used it to perform an asteroid mission by 2016. But the President, while calling for such a flight, actually terminated the programs that would make it possible. Without the skill and experience that actual spacecraft provides, the USA is far too likely to be on a long downhill slide to mediocrity. (Zubrin April 19, 2010)

With the cancellation of the Constellation program, the Space Shuttle, and the ISS in 2024, NASA's morale fell. It became apparent that aerospace insiders who valued human space exploration were extremely frustrated because they did not see these values mirrored in US space policy during this presidency (Jackson April 14, 2010).

However, there were others which applauded the decision to use private aerospace contractors such as Boeing and SpaceX to develop the capability to ferry astronauts to the International Space Station. Others have suggested that a cut back in NASA and other civil space agency expenditures around the world have been key to the renaissance in private enterprise in space and the now forecasted a trillion-dollar space economy on the horizon. There is a strong possibility that this would have never happened without the changes in the NASA space initiatives. Thus, some were able to see a positive side in the midst of the US space policy modifications. Indeed the "NewSpace" revolution began to reverberate around the world.

The Silver Lining

"Silver lining" is a term used when one wants to emphasize the hopeful side of a gloomy or unfortunate situation. There is no doubt that the US space policy under President Obama caused a shift in the growth of American companies such as SpaceX, Blue Origin, and many others. Commercial spaceflight ventures have grown tremendously, and both Orbital ATK (now Northrup Grumman Innovation) and SpaceX are flying commercial resupply missions to the International Space Station with NASA as a customer. In addition, the Commercial Crew Program came into fruition during Obama's presidency.

In September of 2014, NASA awarded contracts to two private space companies—both Boeing and SpaceX to ferry American astronauts to the International Space Station. The third option, the Sierra Nevada Corporation's proposal for the use of their Dream Chaser vehicle, ultimately lost out in the competition. NASA had been paying the Russians to send US astronauts to the ISS on the Soyuz rockets launching from Kazakhstan. Under the Commercial Crew Program, these two private space companies were contracted to fly twelve crewed flights to the ISS (six each) beginning in 2020 or 2021. The name of the space capsule for Boeing is the Starliner, while SpaceX named their spacecraft the Dragon. Nine veteran American astronauts have been chosen to fly these first spaceflights, and their journeys will begin a new era in human spaceflight. NASA has worked alongside Boeing and SpaceX ensuring these companies develop and design reliable and safe transportation for the human crews. If these commercial companies can successfully accomplish NASA's goals, it will set a new precedent for US space policy and allow NASA to focus on further human exploration beyond low Earth orbit and other deep space destinations.

Former Commander of the Space Shuttle, astronaut Chris Ferguson, is the director of crew and mission operations for Boeing's commercial crew program. He has been advising Boeing every step of the way to ensure safety guidelines for the astronauts. "What we really see now is a much greater emphasis on safety," said Ferguson. "We're returning to a full capacity ascent abort system to keep astronauts safe all the way through the profile, and that's something that the Space Shuttle didn't have" (Anna Heiney August 6, 2018).

Both of these companies will usher in a unique chapter in human spaceflight and will bring their twenty-first-century approaches to the creation of the Starliner and the Dragon spacecraft. It will be an amazing achievement for the private space industry. Gwynne Shotwell, Chief Operating Officer of SpaceX stated, "The 7000 women and men of SpaceX understand what a sacred honor this was for us to be a part of this program, and for us to fly NASA astronauts. So, thank you very much, we take it seriously, we won't let you down" (Gwynne Shotwell August 3, 2018).

This next generation of human spaceflight originating in the United States, and new commercial space initiatives around the globe, will undoubtedly generate economic growth in the commercial space business, inspire significant innovation for the private space sector, and solidify our international partners around the world.

References

Armstrong, D. (2003). *NASA-Part 1. The history of skylab.* http://www.nasa.gov/missions/shuttle/f_skylab1.html. 6 Oct 2019.

Augustine, N. R., et al. (2009). *Review of U.S. human space flight plans committee: Seeking a human spaceflight program worthy of a great nation.* September 17, 2019. http://www.nasa.gov/pdf/396093main_HSF_Cmte_FinalReport.pdf. Final Report. NASA.

Bonilla, D. (2009). *Review of U.S. human space flight plans committee.* http://www.nasa.gov/offices/hsf/home/index.html NASA. 12 Jan 2020.

Cernan, E. (April 15, 2010). *Open letter: Neil Armstrong, James Lovell and Eugene Cernan urge Obama not to Forfeit US Progress in Space Exploration.* https://www.theguardian.com/commentisfree/cifamerica/2010/apr/15/obama-nasa-space-neil-armstrong.

Connolly, J. F. (2006). *Constellation program overview.* NASA constellation program office. http://www.nasa.gov/pdf/163092main_constellation_program_overview.pdf. 6 Oct 2019.

Connolly, J. F. *Personal interview.* NASA Johnson Space Center. 4 Nov 2014.

David, L. (February 7, 2011). *New secrets of huge soviet moon rocket revealed.* Space.com.

Dick, S. J. (2008). *Why we explore. The birth of NASA.* http://www.nasa.gov/exploration/whyweexplore/Why_We-29.html. 7 Sept 2019.

Dunbar, B. (2015a). *About project mercury.* http://www.nasa.gov/mission_pages/mercury/missions/manned_flights.html. 12 Oct 2019.

Dunbar, B. (2015b). *Apollo 11 mission overview the eagle has landed*. http://www.nasa.gov/mission_pages/apollo/missions/apollo11.html. 12 Oct 2019.

Ezell, E. C. (1978). The partnership: A history of the apollo-soyuz test project. http://www.hq.nasa.gov/office/pao/History/SP-4209/toc.html. 12 Oct 2019.

Heppenheimer, T. A. (2004). *The space shuttle decision, NASA's search for a reusable space vehicle*. NASA History Office. https://ntrs.nasa.gov/citations/19990056590.

Heiney, A. (2018). https://www.nasa.gov/feature/nasa-assigns-first-crews-to-flycommercial-spacecraft.

Hitt, D. (2011). *What was the gemini program?* NASA educational technology services. http://www.nasa.gov/audience/forstudents/k-4/stories/what-was-gemini-program-k4.html. 12 Oct 2019.

Jackson, D. Obama's NASA policy: The white house vs. Neil Armstrong. *USA Today*. April 14, 2010. http://content.usatoday.con/communities/theoval/post/2010/04/obama-to-talk-policy-after-criticism-by-neil-armstrong/1. 12 Nov 2019.

Logsdon, J. M. (2008). *Ten Presidents and NASA: Richard M. Nixon, 1969–1974*. https://www.nasa.gov/50th/50th_magazine/10presidents.html. 12 Oct 2019.

NASA. (2014). Project Apollo: A retrospective analysis. http://www.nasa.gov/Apollomon/apollo.html. 10 Dec 2019.

Obama, B. (April 15, 2010). *Space exploration in the 21st century*. http://www.nasa.gov/news/media/trans/obama_ksc_trans.html. 12 Oct 2019.

Pelton, J. N., Oslund, J., & Marshall, P. (Eds.). (2004). *Satellite communications: Global change agents*. Mahwah: Lawrence Erlbaum Associates.

Reagan, R. (January 25, 1984). *President Reagan's statement on the international space station*. Excerpts of President Reagan's State of the Union Address. http://history.nasa.gov/reagan84.html. 6 Nov 2019.

Russ, D. "Neil Armstrong writes a letter to Obama, one that perhaps we should all read." Civilian Military Intelligence Group, April 14, 2010.

Shotwell, G. (2018). https://www.nasa.gov/feature/nasa-assigns-first-crews-to-flycommercial-spacecraft.

Skytland, N. (2012). NASA declassification management program: Corona program. msl.jpl.nasa.gov/programs/corona.html. 6 Nov 2019.

Zubrin, R. Obama's Failure to Launch. *New York Daily News*. April 19, 2010. http://www.Marssociety.org/portal/obamas-failure-to-launch/. 12 Oct 2019.

Chapter 4
The Trump Administration Space Policy

The majority of the US population considers humans space exploration a triumph of American spirit and innovation and a key component of establishing global leadership. The Trump Administration and the reactivated National Space Council have created a sense of urgency by pressing for a Moon return within a 5-year period. The Artemis program has been conceived as paving the way back to the Moon and laying a foundation for a mission to Mars (Fig. 4.1).

Certainly, one of the significant changes in US space policy has been the re-establishment of a National Space Council as of June of 2017. This organization is chaired by the Vice President and is comprised of high-ranking members of the president's cabinet and key leaders in the federal government. The members of the Council include the Secretary of State, Secretary of Defense, Secretary of Commerce, Secretary of Transportation, Secretary of Homeland Security, Director of National Intelligence, Director of Office of Management and Budget, National Security Advisor, Homeland Security Advisor, Chairman of the Joint Chiefs of Staff, Secretary of Energy, Assistant to the President for Economic Policy, Assistant to the President for Domestic Policy, and the NASA Administrator.

The National Space Council also incorporates a high-level User's Advisory Group which includes select members from the military, astronaut corps, and civil, commercial, and national security sectors whose experience combined with the National Space Council shares expertise and fosters cooperation across all sectors of space utilization—and significantly guides the strategy for the United States' space endeavors. The Trump Administration plan for American space policy has been outlined in five main policy directives to guarantee that the United States remains a global leader in space exploration, with a prioritized effort to return astronauts to the lunar surface by 2024 and follow-on Mars missions (NASA Space Policy Directive. 1 December 14, 2017).

A. Reneau, *Moon First and Mars Second*, SpringerBriefs in Space
Development, https://doi.org/10.1007/978-3-030-54230-6_4

Fig. 4.1 President signs new Space Policy Directive.

Space Policy Directive 1

By December of 2017, President Trump had signed Space Policy Directive 1, which directs NASA to return astronauts to the Moon. This American-led plan calls for an orchestrated effort between NASA, the private sector, and international partners to "lead an innovative and sustainable program of exploration with commercial and international partners to enable human expansion across the solar system, and to bring back to Earth new knowledge and opportunities. Beginning with missions beyond low earth orbit, the United States will lead the return of humans to the Moon for long term exploration and utilization, followed by human missions to Mars and other destinations."

President Trump stated, "The Space Policy Directive I am signing today will refocus America's space program on human exploration and discovery. It marks the first step in returning American astronauts to the Moon for the first time since 1972, for long term exploration and use. This time, we will not only plant our flag and leave our footprints—we will establish a foundation for an eventual mission to Mars, and perhaps someday, to many worlds beyond" (NASA Space Policy Directive 1 December 14, 2020).

Vice President Mike Pence, Chair of the National Space Council, also shared the enthusiasm and serious focus of the President. He said, "Under President Trump's leadership, America will lead in space once again on all fronts. And as the President has said, space is the next great American frontier—and it is our duty--and our destiny—to settle that frontier with American leadership, courage and values" (NASA Space Policy Directive 1 December 14, 2020).

Much like his predecessor, President Trump initially requested the 2017 budget stay steady for many of the critical components of NASA's human spaceflight programs. These projects included the Space Launch System, or SLS, the super rocket being constructed to take humans to the Moon and Mars. Also included was the Orion crew capsule being developed for missions beyond the Moon, as well as budgetary provisions for the Commercial Crew program. As noted earlier, this Commercial Crew program is designed to use commercial rockets and spacecraft to transport astronauts to and from the International Space Station (ISS). Until 2020, the United States had no way to ferry its own astronauts to the ISS and was forced to purchase seats on the Russian Soyuz rocket at the price of approximately 81 million dollars per seat.

Space Policy Directive 2

Space Policy Directive 2 was issued on May 24, 2018 with a goal of streamlining the regulatory process for private actors in the space industry. In brief it states, "It is the policy of the executive branch to be prudent and responsible when spending taxpayer funds, and to recognize how government actions, including federal regulations, affect private resources. It is therefore important that regulations adopted and enforced by the executive branch promote economic growth, minimize uncertainty for taxpayers, investors, and private industry; protect national security, public safety, and foreign policy interests; and encourage American leadership in space commerce" (NASA Space Policy Directive 2 May 24, 2018).

For many American aerospace companies, the regulatory process was a confusing web to navigate for approval and licensing, and several corporations were beginning to move to countries where there was less hassle. SPD-2 was intended to create a consumer-friendly regulatory environment, minimize red tape, and keep United States' companies operating from American locations. The policy calls for the Secretary of Transportation to transfer this responsibility to the Department of Commerce and create a "one-stop-shop" for commercial spaceflight licenses–although this has never been funded by Congress.

Space Policy Directive 3

In order to address the issue of space object congestion, the President signed Space Policy Directive 3. Currently the Department of Defense tracks over 20,000 objects in space, of which only 2000 are operational satellites. With the implementation of the S-band radar system being installed in the Marshall Islands, the number of tracked objects is expected to increase to about 300,000 objects. The number of space debris objects is expected to continue to increase with the launch of a massive amount of small satellites, largely in LEO constellations in the next 5 years (NASA Space Policy Directive 3 June 18, 2018).

Space debris can be a result of defunct satellites in orbit, spent rocket stages, space collisions, and anti-satellite weapon demonstrations. A collision with just one of these small objects can be extremely life-threatening for our astronauts, as well as current satellites in orbit. In addition, SPD-3 encourages other nations and commercial operators to be responsible for space traffic management, tracking, and disposal. In this regard, it sets forth practical guidelines to avoid on-orbit collisions. Fortunately, most of the operators of the small satellite constellations are planning to remove debris and defunct satellites promptly at the end of their functionality.

Space Policy Directive 4

Initially, the exploration and use of outer space were intended for peaceful purposes, but unfortunately, this realm has increasingly tilted toward being a warfighting domain, despite the provisions of the Outer Space Treaty that calls for the peaceful and non-military uses of outer space. This treaty which is currently in force for the great majority of all countries in the world, states the following in Article IV: "State(s) Parties to the Treaty undertake not to place in orbit around the Earth any objects carrying nuclear weapons of any other kinds of weapons of mass destruction, install such weapons on celestial bodies, or station such weapons in outer space in any other manner. The Moon and other celestial bodies shall be used by all States Parties to the Treaty exclusively for peaceful purposes. The establishment of military bases, installations and fortifications, the testing of any type of weapons and conduct of military maneuvers on celestial bodies is forbidden" (NASA Space Policy Directive 4 Feb. 19, 2019).

The last of the new Space Policy Directives concerns the strategic and potential military uses of outer space. Space Policy Directive 4 directs the Department of Defense to establish a US Space Force which would have become the sixth branch of the US military in order to "beef up" the necessary resources to ensure US national security, guarantee greater protection for space assets, and provide for additional space defense capabilities. This proposal required legislative approval and would have enabled the Department of Defense to create and equip this new organization to protect and secure any and all threats in the space domain.

After much legislative opposition, on December 20, 2019, the National Defense Authorization Act for 2020 was passed which redirected the former Air Force Space Command to become the US Space Force (USSF) as a space operations branch of the US Air Force. The US Space Force's primary focus is to "organize, train, and equip space forces in order to protect US and allied interests in space and to provide space capabilities to the joint force. Its responsibilities include developing military space professionals, acquiring military space systems, maturing the military doctrine for space power, and organizing space forces to present to the Combatant Commands."

Critical in this regard is greater resilience for military satellites that provide secure communications, navigation, early warning systems for missile attacks, and precise munitions targeting. It is regrettable that outer space has become a warring domain, but this is the new world we all live in.

The Artemis Program

NASA, the White House, and the National Space Council, are currently implementing a plan to return to the Moon by 2024–2025. The defined goal at this time is directed toward an effort to set up a permanent presence on the lunar surface. The name of this project is "Artemis" inspired after the goddess of the Moon in Greek mythology and also the twin sister of Apollo. It is no accident that the name of this program is female, as NASA will return American astronauts to the lunar surface, guaranteeing this time that the first woman will step foot on the Moon. The new landing point will be the South Pole, to explore the benefits of possible use of the large amount of water-ice located there. This water, if accessible, could be a tremendous in situ resource that could enable humans to stay for longer periods on the Moon and, with the right processing capabilities, could also possibly be converted into rocket fuel.

The future of the Artemis program will require not only technological development but will also likely pose regulatory and legal issues as well. The issue could be whether the Outer Space Treaty of 1967 and the subsequent Moon Treaty restricts or might even prevent the removal of resources from the Moon without some form of international agreement to share in the benefits derived from any such materials, such as water/ice. The Moon Treaty has not been ratified by any nation that engages in human space flight since its creation in 1979 and should have no relevancy in international law. National legislation passed in the United States, United Arab Emirates, and Luxembourg within the last 3 years have sought to establish the right of these countries and national companies to derive space resources from the Moon and celestial bodies. To date there has not been any definitive interpretation of the Outer Space Treaty and the Moon Agreement as to the conditions under which space mining might be undertaken. As efforts like the Artemis program as well as private initiatives pursue their goals for a permanent presence on the Moon, it will be increasingly important to resolve the legal and regulatory issues associated with use of space resources.

The original plan for the next American astronaut Moon landing had been set by NASA for the achievable and relatively comfortable goal of 2028. However, the latest presidential mandate has now directed NASA to move the landing date forward to 2024. Consequentially, this caused Vice President Pence to issue a daring challenge which was distinct and direct: "Failure to achieve our goal to return an American astronaut to the Moon in the next five years is not an option." His words shook up the complacency of NASA and signaled a wakeup call to the agency's leaders, employees, and contractors. It would no longer be business as usual with repeated cost overruns, unending delays and embedded bureaucracy. "We will call on NASA not just to adopt new policies, but to embrace a new mindset. That begins with setting bold goals and staying on schedule. NASA must transform itself into a leaner, more accountable, and more agile organization. If NASA is not currently capable of landing American astronauts on the Moon in 5 years, we need to change the organization, not the mission" (Newt Gingrich March 30, 2019).

The goals of Artemis and NASA's Moon mission are well defined. This reinvigoration of NASA will push America's space exploration boundaries and will also include collaboration with commercial companies and NASA's international partners. If Artemis can come to fruition, the United States will be able to create a sustainable human presence, generate new scientific discoveries, as well as prove advancing technologies. These advancements combined will lay the foundation for a human journey to Mars—and beyond.

Architecture of Artemis (Fig. 4.2)

The first phase of the Artemis 1 mission is scheduled to launch from Kennedy Space Center in 2021 which has been delayed from its initial launch in late 2020. This first flight will be an uncrewed test flight of the Space Launch System (SLS) rocket integrated with the Orion human-rated space capsule. This mission will have a duration of approximately 3 weeks.

The test flight will travel thousands of miles/kilometers beyond the Moon. If Artemis I is successful, then the next phase will be Artemis 2. This second test flight is scheduled to occur in 2022 or 2023 and will include an integrated SLS and Orion which will ferry American astronauts to lunar orbit in a trial run. The architecture of the SLS rocket is such that there is to be a separate configuration for crew launches (known as the Block 1B Crew to be followed by Block 2B Crew) and then there is a different configuration (Block 1B Cargo and Block 2B Cargo) for carrying cargo (Artemis 1 Feb 3, 2020) (see Fig. 4.3 for one of the SLS cargo configurations).

The Artemis 3 mission is then planned to follow in 2024. This operational flight will include human landing on the surface of the Moon with astronauts. The crew would make their landing on the moon in a human lunar lander, which may be built by American companies with NASA oversight.

Lunar Orbital Platform Gateway

The Artemis program is planning to be supported by the Lunar Orbital Platform Gateway (LOP-G); although with recent personnel changes, this program is still being considered and may well be eliminated. During NASA's announcement about the postponement of Lunar Gateway, Doug Loverro former head of Human Exploration and Operations Mission Directorate (HEOMD) said, "Not a single

Fig. 4.2 Space Launch System. (Graphic courtesy of NASA)

international partner is ready to do anything on Gateway until 2026" (Doug Loverro March 13, 2020).

The change of plan and new schedule will save time and reduce costs and allow NASA to pursue its ambitious plan of a 2024 Moon landing. By streamlining this plan, the United States does not need to worry about creating new technology or to test new space systems which, in turn, reduces a significant amount of risk.

However, until these policy decisions are finalized, I will briefly describe the Lunar Gateway plan. This is a space station that is known as the "Lunar Gateway" for short. It is to be placed in lunar orbit in stages. This first part of the lunar space station will serve as a "staging" hub for robotic spacecraft, telescopes, scientific equipment, and a refueling depot and will also be a short-term habitation module for up to four astronauts. The Gateway will be developed in cooperation with NASA's international and commercial space partners and will consist of four modules.

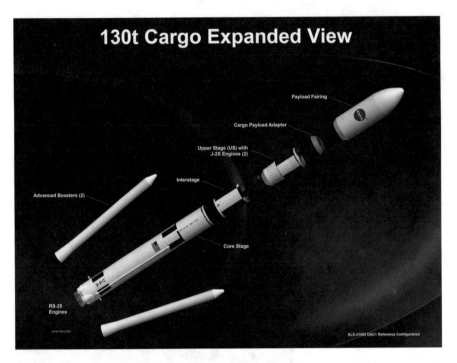

Fig. 4.3 The Space Launch System 130t Cargo Version. (Graphic courtesy of NASA)

The first module will be launched and is intended to be the solar-electric Power and Propulsion Element (PPE). The PPE will serve as the command and communications center and enable orbital transfers, as well as a space tug for transfers to and from the Moon. The second element will be named the Minimum Habitation Module (MHM). The function of this module is just as it sounds—a small version of a space station. It will include life support systems integrated with the Orion spacecraft and can maintain four astronauts for up to 30 days.

The third module to be launched will be called the Gateway Logistics Module and will be used for logistics, as well as resupply and refueling of the lunar space station. In addition, the logistics module will include the CanadArm 3. This is to be a highly advanced robotic arm built by the Canadian Space Agency. The final stage will be the Gateway Airlock Module and will be used for human missions performed outside the station called EVA's or extravehicular activities, although the Lunar Gateway has no definite launch date at this time.

The Lunar Gateway will be used to study many areas of science including astrophysics, planetary science, medicine, biology, botany, heliophysics, human health, and more. Scientific experiments are scheduled to begin once NASA has accomplished its first priority—landing astronauts on the Moon (see Figs. 4.3 and 4.4).

Artemis has reenergized the United States' space program and laid out a clear and focused plan which will broaden the boundaries for human presence beyond low Earth orbit. The current administration has provided a well thought out vision

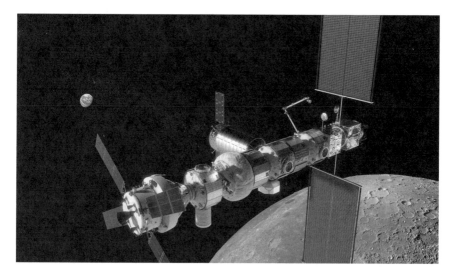

Fig. 4.4 The complete Lunar Gateway shown in lunar orbit and the Earth in the distance. (Graphic courtesy of NASA)

for NASA which will effectively include its international and private partners. This effort to create sustainable human and robotic exploration will greatly benefit Earth and provide economic growth and scientific knowledge to create a better life on our planet—and a sustainable existence beyond.

References

Artemis 1 Overview, NASA. https://www.nasa.gov/artemis-1. Last accessed 3 Feb 2020.

Gingrich, N. (March 30, 2019). https://www.gingrich360.com/2019/03/making-the-trump-pence-space-challenge-a-reality/. Last accessed 18 Feb 2020.

Loverro, D. (March 13, 2020). https://spacepolicyonline.com/news/gateway-no-longer-mandatory-for-2024-moon-landing/. Last accessed 27 Mar 2020.

NASA. *President signs new space policy directive*. https://www.nasa.gov/press-release/new-space-policy-directive-calls-for-human-expansion-across-solar-system. Last accessed 20 Feb 2020.

US Space Policy Directive-1. *Space News*. December 14, 2017. https://spacenews.com/tag/space-policy-directive-1/. Last accessed 3 Feb 2020.

US Space Policy Directive-2. (May 24, 2018). https://www.whitehouse.gov/presidential-actions/space-policy-directive-2-streamlining-regulations-commercial-use-space/. Last accessed 3 Feb 2020.

US Space Policy Directive-3. (June 18, 2018). https://www.whitehouse.gov/presidential-actions/space-policy-directive-3-national-space-traffic-management-policy/. Last accessed 3 Feb 2020.

US Space Policy Directive-4. (February 19, 2019). https://www.whitehouse.gov/presidential-actions/text-space-policy-directive-4-establishment-united-states-space-force/. 9. Last accessed 13 Feb 2020.

Chapter 5
Moon and Mars: A Comparison of the Two

Fig. 5.1 The comparative size of Earth and Mars. (Graphic courtesy of NASA)

Even though Elon Musk envisions humans living and working on Mars, NASA and its partners are preparing to undertake the quest of returning men and women astronauts to the Moon. All of the human beings on planet Earth will benefit as we embark on these adventures and explore other celestial bodies. Our universe is awe-inspiring—and when we look up into the night sky, it is impossible to deny how grand and impressive it is! This fascination is what motivates us to stand behind our national space agencies as they conceive and construct the necessary architecture to

A. Reneau, *Moon First and Mars Second*, SpringerBriefs in Space
Development, https://doi.org/10.1007/978-3-030-54230-6_5

venture into the unknown. Yes, we have been to the Moon, but we envision and dream of going to Mars. But which destination is the best choice at this time in our history? This chapter will seek for an answer by comparing the two destinations and the unique set of challenges that should be considered for each one. We will need to examine what natural resources exist, proximity to Earth, soil content, atmosphere and weather, temperature, gravitational issues, and many other components to make a sound and logical conclusion (Fig. 5.1).

The US-led Artemis program (which at this time includes the Lunar Gateway) has a number of international and commercial partners. As stated above, the purpose of this program seeks a rapid return of astronauts to the Moon as a vital step in the process of sending human explorers to Mars. As the pioneering SpaceX aspires to go directly to the red planet, American public opinion is not strong on either option. But, they do see a Mars human exploration quest as not only a feat to be accomplished, but also a worthwhile goal. But what is the correct answer? Perhaps there is no simple right or wrong answer. The right question would be—what is currently technically and financially feasible in the geopolitical and economic climate? There are several differences between these two worthy destinations, and a comparison between them should reveal possible clues as to which is the best choice—for now.

Distance

A journey to the Moon is approximately 3 days away at a distance of 238,000 miles (or about 400,000 kilometers) and humans can "come and go" without great difficulty and relatively safely—compared to a mission to Mars. This is because current rocket technology is advanced enough to travel back and forth at this close distance to Earth. Also, the proximity of the Moon from our planet saves a significant amount in fuel costs. With our current rockets, we can begin transporting all the supplies needed in advance for humans to live and work on the Moon.

The mighty Saturn V rocket, as big as it was, was only large enough to transport three astronauts to lunar orbit in a spacecraft with the interior size of a small minivan (10.6 feet high by 12.8 feet wide/3.2 m by 3.9 m). Imagine trying to create a spacecraft with living quarters for a six-month trip to Mars for at least four astronauts—and the size of the rocket fully laden with the amount of fuel it would require. While it is true that we do have rockets that can land rovers on Mars, it requires a tremendous amount of tonnage to take the supplies necessary to take astronauts and all the support systems for a safe journey. Following the time spent on the red planet's surface, you would have to have a large enough rocket—and fuel—to launch the Martian astronauts back home again (Ridley October 18, 2017).

Gravity

Gravity is a force that draws one object toward another, just as our sun has a gravitational pull on all of the planets in our solar system. This unseen force keeps our feet firmly planted on the ground, holds down our atmosphere, and regulates the air we need to breathe. Even though scientists cannot tell us exactly what gravity is, it is a serious factor to be considered when it comes to human space exploration.

Consider the Moon which has about 1/6 of Earth's gravity—or about 17%. This would mean if you weighed 150 pounds/68 kilograms on Earth, you would weigh 25 lbs/11 kg while standing on the Moon. This is not ideal for sustaining bone density and strong muscles in the human body. Mars, on the other hand, has 38% of the gravity of Earth, which is much better physiologically for humans. If you weighed 150 lbs/68 kg on Earth, your Mars weight would be 57 lbs/26 kg. Many experts believe that people could actually evolve to cope with Mars gravity for extended periods of time. However, the Moon's lesser gravitational factor would cause quicker deterioration of bone and muscle mass, making it more difficult for long-term habitation. But, the advantage of low gravity on the Moon allows for it to be an efficient launch site for rockets heading into deep space travel, as it requires much less rocket fuel to escape the gravity well (or gravity dimple) of the Moon. Mars, however, requires much more fuel to launch our astronauts back to the Earth—or destinations further into our solar system.

Atmosphere

The atmospherical difference between the Moon and Mars is quite significant and is a major factor when determining travel to either of these locations. The Moon has little or no atmosphere or weather and does not offer any protection from cosmic radiation. Whereas explorers might be able to live aboveground on Mars, they would definitely have to live under the Moon in caves or lava tubes or in heavy radiation protected aboveground dwellings. Also, because there is no atmosphere on the Moon, there is no wind or weather. This makes the temperature very hot on the sun-drenched side of the Moon and extremely cold on lunar "nights."

Mars has a thin atmosphere which allows for some protection from solar particles and cosmic radiation. The air consists mostly of carbon dioxide which is actually good for plant life, but, obviously, not so good for Martian dwellers and astronauts. The light atmosphere does create wind and weather which balances out day and night temperatures slightly better than the Moon, and results in the presence of wind as well. This wind can cause violent dust storms, from which astronauts would have to take cover, but the wind could be harnessed for energy. People could live aboveground in pressurized habitats which would be much more convenient than underground Moon dwellings. Mars is definitely more people-friendly and, as a result, is a much better choice for extended stays.

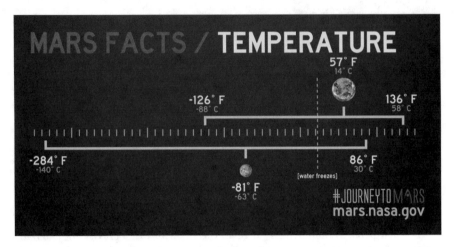

Fig. 5.2 The comparative temperatures of Mars and Earth. (Graphic courtesy of NASA)

Temperature

Moon explorers would encounter extremely high temperatures during a Moon day (250 °F/121 °C) and very frigid temperatures during a Moon night (−380 °F/−229 °C). The drastic temperature fluctuation of approximately 600 degrees Fahrenheit has a dramatic effect on astronauts and their equipment. This contrast of temperatures between the 14 day-night cycles makes it very challenging to create durable life-support systems, transportation vehicles, robots, batteries, and space suits, among other things. Think of how difficult it would be if you lived on the hot beaches on Cancun, Mexico, and were suddenly living in Antarctica 14 days later. This deviation in temperature is extremely problematic (Ridley October 18, 2017) (Fig. 5.2).

A Mars day is a bit more normal for humans with temperatures reaching 80 °F/27 °C during the day but dropping to nearly −200 °F/−129 °C at night—at the poles. This temperature change is only about a 300 degree difference, which could be more manageable for Martian explorers and their necessary equipment and systems.

Day and Night Cycles

The comparison between the length of days and nights on Moon and Mars is vastly dissimilar. The amount of time the Moon takes to complete a turn on its axis is almost 28 Earth days. The "days" last about 14 days and include steady sunlight, followed by 14 days of extreme darkness. The 28-day dark and light lunar cycle makes it difficult to grow life-sustaining crops and, as aforementioned, is challenging for humans with the excessive temperature changes. Growing crops could possibly be accomplished with specialized greenhouses which allow in the sunlight, blocking harmful radiation, and regulating temperature for optimum plant growth and reproduction. One way around this 14-day problem is to establish an outpost at the Moon's poles where sunlight or darkness is constant.

Mars has a more Earth-like day and night cycle as a Mars day is 24 hours and 39 minutes long. The tilt of the red planet's axis is within half of a degree of our planet which results in seasons. Plants would still need to be grown indoors in a greenhouse setting due to Mars' thin atmosphere, daily temperature changes, and solar radiation. Another viable option could be hydroponics (growth of plants without soil).

Soil and Regolith

If humans are going to live long term on other celestial bodies, growing food will be essential. Consider the recent Hollywood movie "The Martian" where the astronaut, played by Matt Damon, found a way to grow potatoes to survive. The Moon's crust and associated dust, referred to as regolith, is a serious hazard for astronauts and their equipment. But research scientists using NASA-approved artificial regolith have conducted plant growth experiments to test the possibility of agriculture endeavors on the moon. This man-made regolith was mixed with other organic materials to test the growth of ten types of plants for human consumption. These crops included tomatoes, peas, chives, quinoa, radishes, rye, onions, spinach, and cress. The plants all thrived with the exception of the spinach.

The same type of experiment was conducted using NASA-approved simulated Martian soil which was also mixed with organic compounds. The plants did much better in the Martian soil mix than the Moon mixture. Scientists were able to grow actual tomatoes and radishes, etc., and seeds were harvested to produce a second crop.

Although this is a tremendous outcome, there is still much research to do before successful agriculture exists on these two bodies. In the case of this experiment, only soil was considered, not the extreme temperatures, or the day-night cycles. However, this is positive progress and another extraordinary step toward "off-Earth" living (Nield October 18, 2019).

Communication and Travel

Because of the excessive distance between Mars and Earth, it takes radio signals a long time to be relayed. This delay can't be improved by more powerful radios or faster computers. The communication delay is between 4 and 24 minutes depending upon Mars proximity to Earth. If Mars is on the Earth side of the sun, it is the lesser time frame, but if it is on the opposite side of the sun, then it will be nearly 24 minutes. The signals can only travel at the speed of light, which is the fastest anything can travel in our known universe.

Light travels at the speed of 186,000 miles per second, and at this rate, one could circle the Earth 7.5 times or be on the Moon in less than 2 seconds. Perhaps this puts into perspective how far away Mars truly is from our home planet! This is a serious challenge for exploration to Mars because the delay makes it difficult for real-time communication. In fact, NASA refers to two different times when dealing with spacecraft event times: SCET (spacecraft event time) which is UTC time on board the actual spacecraft, and ERT (earth received time) or the time a signal was received on Earth at a deep space network (DSN) terminal.

Upon first assessment, it appears Mars is a more people-friendly planet and a desirable location for humans. But the obstacle of distance—for now—is the largest "deal killer." Mars is 6–8 months away, via today's rocket systems, at a distance of approximately 46 million miles (nearly 74 million kilometers), but only when the Earth and Mars are closest together with an aligned orbit. Once a crew arrives at Mars, they would need to be prepared to stay for probably at least a year or more until Earth and Mars are again in orbital locations that are comparatively close. In short, there is no chance of a quick rescue and return operation from Earth if the crew traveling to Mars were to encounter a problem. This challenge alone generates problems of both physiological and psychological stress. The gravitational effects will result in bone and muscle loss, and solar and galactic radiation can cause cancer since the atmosphere and lack of a protective magnetic shielding are not able to provide enough protection against these harmful effects. Furthermore, the cumulative amount and weight of food, water, fuel, and medicine needed to complete a successful mission seems prohibitive. This also does not include the extremely challenging entry, descent, and landing (EDL) system, which must be capable of delivering at least ten times the mass and volume of our current robotic missions to Mars (Herath April 18, 2011).

With the close distance of the Moon and the discovery of millions of tons of water-ice in the polar regions, the Moon may be the best choice for now. This frozen water, when processed and purified, may be life-giving to humans and can be used as a natural resource for rocket fuel if the water were split into its hydrogen and oxygen components. Because the Moon is only three days away, if something were to go wrong with the crew, equipment, or spacecraft, a rescue and return is feasible. Radio signals with only a two-second delay is technology that is already in place for robotic and astronaut communication. A lunar landing for the many comparisons brought forth in this chapter is a more reasonable and safe choice for the 2020s.

References

Herath, A. K. "Why is it So Hard to Travel to Mars?" *Astrobiology Magazine,* April 18, 2011. http://m.space.com/11417-Mars-mission-space-travel-challenges.html. Last accessed 10 Feb 2020.

Nield, D. (October 18, 2019). *Peas, quinoa and 7 other crops grown successfully in soil equivalent to Moon and Mars.* https://www.sciencealert.com/experiments-show-soil-from-the-moon-and-mars-could-support-crops. Last accessed 13 Feb 2020.

Ridley, A. (October 18, 2017). *Is it better to live on the moon or on Mars? A scientific investigation.* https://qz.com/1105031/should-humans-colonize-mars-or-the-moon-a-scientific-investigation/. Last accessed 11 Feb 2020.

Chapter 6
A Psychological and Physiological Perspective

Over the past 40 years, NASA has successfully landed a series of robots on Mars, beginning with Viking 1 in 1976. Most recently, the landing of Curiosity in 2012 at Gale Crater, the seventh robotic landing on Mars, and Insight in November of 2018 have inspired the American people and captured their attention.

In May 2012, NASA put together a study group that rendered a tentative goal of a human mission to Mars by 2033. There is a hefty price tag that goes along with this goal, as well a vexing set of challenges. In addition to technical and political obstacles, and unlike the earlier seven robots, humans traveling to Mars will need food, water, protective shelter, medical supplies, entertainment, friendship, and, yes, a return ticket back to Earth. This chapter covers a unique set of problems that must be solved for a successful manned trip to and from Mars. By comparison, there are fewer challenges related to a journey to the Moon.

Psychological Effects

Separation, long-term isolation, and the dynamics of living with fellow astronauts for an extended period of time are some of the challenges that must be addressed. While the Apollo missions lasted a week, the crews could still capture views of Earth, and they knew they were only 3 days away. Although Space Station astronauts rotate to and from the ISS about every 6 months, they can look out ISS windows, see their familiar home planet, and know they are a seeming "stone's throw" from an expected safe return. This is not so with a trip to Mars. Every 26 months there is a brief optimal departure period from Earth, and the round trip, combined with the length of stay, would be expected to last two to three years. As the spacecraft moves toward Mars, Earth becomes a small dot and eventually fades into the vast universe of billions of twinkling stars.

© The Editor(s) (if applicable) and The Author(s), under exclusive license to
Springer Nature Switzerland AG 2021
A. Reneau, *Moon First and Mars Second*, SpringerBriefs in Space
Development, https://doi.org/10.1007/978-3-030-54230-6_6

Astronauts are not going to Mars to plant a flag, pick up some rocks on the surface, perform a few experiments, and leave like they did with Apollo. They are going to spend over a year traveling to and from Mars, and then stay for an extended period of time getting to know the secrets that Mars has yet to reveal. In preparation for the psychological effects that astronauts might encounter, a group of astronauts from Russia, the European Space Agency, and China participated in the "Mars 500 Experiment" from 2007 to 2011 in Moscow. The study simulated a 520-day round trip to Mars in which volunteers lived and worked in a mock mission environment. The experiment generated helpful data on the psychological and social effects of people placed in a long-term, cramped living situation. During the study, communication with the outside world had a realistic time delay of 25 minutes, and there was a limited supply of food and other consumables. Some of the crew members reported trouble sleeping and exercising and would isolate themselves from each other in a type of hibernation. But there were no reports of conflicts, and any difficulty the crew encountered, they resolved together as a team. Overall, the crew members were friendly with each other, and cultural and language differences did not create any significant problems. However, the effects of cosmic radiation and weightlessness were not able to be factored into this experiment.

Of course, there is another option that has been suggested for a Mars mission. This is the idea of sending settlers on a one-way trip to Mars as pioneers of the red planet. This initiative is known as "Mars One." This initiative has gained a large amount of publicity, but it is not generally considered a viable plan. The Mars One website describes their objective as follows:

> Mars One aims to establish a permanent human settlement on Mars. Several unmanned missions will be completed, establishing a habitable settlement before carefully selected and trained crews will depart to Mars. Funding and implementing this plan will not be easy; it will be hard. The Mars One Team, with its advisers and with established aerospace companies, will evaluate and mitigate risks and identify and overcome difficulties step by step. (Mars One Feb 3, 2020)

Physiological Effects

Even though space agencies have been launching astronauts into space for over 50 years, we still do not understand all of the adverse effects that space travel has on the human body. A few of these challenges include exposure to radiation and weightlessness, which can lead to cancer, bone loss, muscle atrophy, vision impairment, heart and circulatory issues, and possible brain damage.

NASA recently conducted a study on astronauts Scott Kelly and his twin brother Mark. The twins agreed to a year-long study (actually 340 days) assessing the impact of long-duration space travel and the human body's reaction to exposure to weightlessness and radiation. Scott Kelly returned from space on March 1, 2016, and the testing continued on both him and his brother Mark, who remained on Earth. The loss of muscle and bone, vision problems, as well as motion and balance will be tested and compared between the twins. The study will help NASA prepare to take humans farther into deep space (Dunbar January 19, 2016).

But there are major differences when undertaking a trip to Mars because a long-duration mission into deep space involves exposure to a different type of radiation. Scott Kelly was in LEO during his mission, and Earth's magnetic field protected the astronauts from the more severe radiation exposure that crews would encounter on a trip into deep space. An astronaut on the ISS encounters about 20 times the amount of radiation compared to Earth. But a journey to the red planet increases radiation 300 times the normal exposure for a human being (Yuhas March 5, 2016).

Beyond LEO, humans will encounter galactic cosmic rays and solar particle events. NASA scientists do not have sufficient knowledge about radiation in space, and they are hesitant to predict the effects on a crew as it hits the spacecraft and would eventually threaten the astronauts during their stay on Mars. According to Brett Drake, former Deputy Chief Architect for NASA's Human Spaceflight Architecture Team, NASA could reduce exposure to normal background radiation in space by building shielding into the spacecraft and the Mars habitats. Drake thinks NASA needs an improved method of predicting life-threatening solar flares, which spew extremely high doses of radiation, so astronauts can retreat to special storm shelters when the need arises (Drake November–December, 2012).

A radiation assessment detector, which was carried along with Curiosity to Mars, was operational during the transit from Earth to Mars. It was determined that if humans had been involved in the journey, their risk of cancer would increase by five percent. Unfortunately, this is higher than NASA's limits for an astronaut. Radiation in deep space can be very damaging as it leaves a number of medical issues in the human body over a lifetime (see Fig. 6.1).

Fig. 6.1 The Curiosity Mars Rover on the red planet. (Graphic courtesy of NASA/JPL)

Long-term exposure can lead to cataracts, as reported by 36 former Apollo astronauts who were part of high-radiation missions. On Earth, cataract surgery is a relatively common procedure, but such surgery would be impossible to perform during a mission to Mars. Also, there is a common problem with vision impairment, as experienced on the ISS. Bob Thirsk, a Canadian astronaut who holds the Canadian record for the longest space flight (187 days) and the most time spent in space (204 days), lost a significant amount of visual acuity. This situation is not unusual among astronauts who have lengthy missions; unfortunately, the damage is permanent (Seedhouse 2016).

Another problem encountered by ISS astronauts is bone loss and muscle atrophy. On average, astronauts on the ISS lose about 1.7% of outer bone mass and 2.5% inner bone mass per month during lengthy stays. Even after a year of rehabilitation, they still may have significant bone loss. Despite vigorous daily exercise while in space, muscle atrophy sets in. When these healthy astronauts return to Earth, they can hardly stand or walk and must be undergo rehabilitation (Seedhouse 2016).

If only a few of these psychological and physiological issues occur, the crew of a Mars mission will be weakened upon entry, descent and landing to the Mars surface. Imagine how difficult it would be for the crew to execute their mission there. These are just a few of the psychological and physiological challenges that must be solved before missions to Mars are undertaken. This is why "dress rehearsals" in cislunar space and on the surface of the Moon are critical for enabling a successful journey to Mars.

References

Drake, B. (2012). The deferred dreams of Mars. *MIT Technology Review, 13*(05), 10.

Dunbar, B. (January 19, 2016). *Twins study*. NASA. https://www.nasa.gov/feature/nasa-s-twins-study-results-published-in-science. Last accessed 12 Jan 2019.

Mars One. (2020). http://www.mars-one.com.

Seedhouse, E. (2016). *Mars via the Moon: The next giant leap*. Cham: Springer Praxis Books.

Yuhas, A. Marathon space flight just the start for Scott Kelly, Walking Science Experiment. *Guardian*, March 5, 2016.

Chapter 7
Technological Challenges

There are a number of key technologies both needed and planned by NASA to explore destinations in deep space. These include the Moon, Mars, asteroids, and the moons of Jupiter. When NASA dictates a destination, it drives the development of key technologies and capabilities for US and other space explorers. NASA focuses on these technologies in order to develop equipment that is needed to build their ability to explore a variety of destinations. The architecture includes transportation systems, mission operations, habitation structures, protective systems for human physiological and psychological health, and destination systems that create an interrelated and evolving infrastructure. As these systems are matured and perfected, it can guarantee a seamless transition from low Earth orbit to the Moon, Mars, and other destinations.

With ambitious goals to go to Mars and perhaps beyond, the technological challenges are very real. The Moon could possibly be the best location for a technological proving ground. If we master the Moon, both in cislunar space and on its surface, that will certify the technologies needed and become a logical staging area for future deep-space exploration. NASA will have opportunities to develop safe operations that will support decades of future missions while remaining in close proximity to Earth. This strategy, it is hoped, will open up the pathways to Mars. The Moon is also an affordable and sustainable destination for research, exploration, and other beneficial activities. These activities, manned and unmanned, will span over many years to come, not only for the United States but also for India, China, Europe, Japan, Canada, Russia, and several other countries.

NASA has identified a list of technologies that are essential to exploring beyond LEO and advance human presence in Earth's solar system. One of the technologies needed is a transportation system beyond low Earth orbit. Included in this are ground operations (facilities for launching spacecraft from Earth), the SLS heavy-launch vehicle, and the Orion crew capsule. Deep-space missions will need to develop a Mars transit vehicle with high-efficiency in-space propulsion and power, protection from radiation, optical communication, and deep-space navigation and rendezvous.

A. Reneau, *Moon First and Mars Second*, SpringerBriefs in Space
Development, https://doi.org/10.1007/978-3-030-54230-6_7

Upon arrival, there must be entry, descent, and landing (EDL) systems capable of delivering far greater mass than the present robotic missions. Surface power generators will be essential for energy production sufficient to meet the requirements associated with human space destinations. Astronauts will need long-duration habitation modules that include life-support systems, radiation safety, protection from the climate, and medical assistance for crew health. Currently, the International Space Station (ISS) is Earth-reliant and dependent on resupply flights for consumables as well as propulsive systems to maintain orbit. Because longer missions could last 1–2 years, astronauts will need to be self-sufficient and Earth-independent. In situ resource utilization is a required development, as well as comfortable EVA spacesuits and sustainable food and water systems. Also necessary are mobile exploration vehicles and eventually ascent propulsion to return back to Earth. Of course, this does not encompass all of the necessary technologies, but they are representative of those that will have to be created and refined into mature systems over time (NASA May 29, 2014).

Missions to the Moon will be useful in preparing for longer journeys in deep space. Such new capabilities will be helpful in successfully demonstrating viable systems that are independent from Earth. Returning to the Moon is thus, in fact, a cornerstone of longer-term deep-space human exploration. The creation and development of Orion and SLS are well under way, but there are still several gaps that need to be addressed as the necessary technologies are developed, modified, and integrated into NASA's exploration goals. One of the key questions is to what extent other partnering nations and private space enterprises can be responsible for developing these new technologies and systems (NASA June 7, 2012).

On February 2016, the House Committee on Science, Space, and Technology's Subcommittee on Space held a hearing entitled "Charting a Course: Expert Perspectives on NASA's Human Space Exploration Proposals." Several expert witnesses were present and testified. They included Paul Spudis, Senior Scientist of the Lunar Planetary Institute; Tom Young, former Director of NASA's Goddard Space Flight Center; and John C. Sommerer, former Chief Technology Officer at John Hopkins University Applied Physics Laboratory and former Chair of the Technical Panel for the "Pathways to Explorations Report," which is part of the National Academy of Sciences (NAS). The Academy conducts studies for the federal government and is comprised of experienced experts in all areas of science and technology including members who were involved in the US space program dating back to the early years of Mercury, Gemini, and Apollo. This non-profit, government-mandated committee is helping NASA determine its technology roadmap. In his testimony, Sommerer stated that the technological demands of a crewed Mars mission are very challenging and that there is a huge gap in current capabilities and funding (National Academy of Sciences, "Pathways to Exploration" 2014).

The "Pathways to Explorations Report" from the National Academy of Sciences Report lists six key principles that NASA should adopt in its future human exploration programs in the figure below. It also listed 15 key technologies that would be critical to extending human exploration to Mars (Fig. 7.1).

NASA should adopt the following Pathway Principles:

I. Commit to designing, maintaining, and pursuing the execution of an exploration pathway beyond low Earth orbit toward a clear horizon goal that addresses the "enduring questions" for human spaceflight.

II. Engage international space agencies early in the design and development of the pathway on the basis of their ability and willingness to contribute.

III. Define steps on the pathway that foster sustainability and maintain progress on achieving the pathway's long-term goal of reaching the horizon destination.

IV. Seek continuously to engage new partners that can solve technical or programmatic impediments to progress.

V. Create a risk -mitigation plan to sustain the selected pathway when unforeseen technical or budgetary.

VI. Establish exploration pathway characteristics that maximize the overall scientific, cultural, economic, political, and inspirational benefits without sacrificing progress toward the long-term goal, namely:

a. The horizon and intermediate destinationshave profound scientific, cultural, economic, inspirational, or geopolitical benefits that justify public investment.

b. The sequence of missions and destinations permits stakeholders, including taxpayers, to see progress and to develop confidence in NASA's ability to execute the pathway.

c. The pathway is characterized by logical feed-forward of technical capabilities.

d. The pathway minimizes the use of dead-end mission elements that do not contribute to later destinations on the pathway.

e. The pathway is affordable without incurring unacceptable development risk;

f. The pathway supports, in the context of available budget, an operational tempo that ensures retention of critical technical capability, proficiency of operators, and effective use of infrastructure.

Fig. 7.1 Pathway Principles to Human Exploration recommended by the National Academy of Sciences. (Graphic courtesy of the US National Academy of Sciences)

This report also presented a detailed matrix for technological development. This was presented as a matrix that listed the technologies to be developed, the challenges to be overcome, and the relative difficulties of these challenges to be met. The result was some 60 different assessments of the difficulties to be met (Pathways to Exploration June 2014).

Eighteen of those intersections were rated green, meaning that progress can be expected with minimal risk. Twenty-four intersections were rated yellow, indicating significantly higher risk. Eighteen of the intersections were rated red, indicating such hurdles as "no technical solution known" and that no such systems have ever been developed.

In commenting on the pathway forward in space exploration during the Hearings of the House of Representative's Subcommittee on Space, Dr. John Sommerer stated:

One of the jobs of the Committee is to establish a pathway for human exploration into deep space. Understandably, the committee suggested that there are only a few credible destinations for humans in the solar system, given our technical knowledge, economic limitations and limits associated with human physiology. They suggested a plan and advocated that the U.S. needs to quit changing its mind. At a minimum, we should agree on a pathway that is satisfying to the public, even if it does not lead to Mars in the foreseeable future. A pathway that includes the surface of the Moon is one obvious possibility. (Sommerer February 3, 2016)

Nuclear Propulsion: The Game Changer

The most current and efficient way to journey into space is by the use of chemical rockets, and a method of travel and orbit-transfer for spacecraft is known as the "Hohmann transfer." Chemical rockets use a highly flammable fuel (usually liquid oxygen and hydrogen) as the propellant. As the propellant is ignited inside the combustion chamber, hot gases are produced by the chemical reaction. These rapidly expanding hot gases are then forced out of the chamber through a narrow rocket nozzle to produce extreme thrust (Foust May 22, 2019).

In the Hohmann transfer method, a rocket would launch from Earth by chemically fueled propulsion and proceed into low Earth orbit. At a specifically calculated moment, the rocket would fire again, sending the rocket into a different elliptical orbit which would eventually intersect with a Mars orbit. The purpose of this type of space travel is to use the least amount of energy when traveling from one planetary body to another. Although this is a very efficient way of travel, its disadvantages are that it takes way too much time. For instance, on a mission to Mars, the Hohmann transfer would require that Earth and Mars be lined up properly for a launch window. The windows of possibility for launch only occur every 26 months. From launch to arrival at Mars, it would take astronauts between 7 and 9 months of travel time. The negative is that during this time, the astronauts would need a tremendous amount of resources, i.e., air, water, food, and fuel; and of critical importance, the exposure to cosmic radiation would be extremely high. Not to mention, there would be an equal amount of resources needed for the return journey as well (Hadhazy December 22, 2014).

But there is a new form of propulsion arising once again, known as nuclear thermal propulsion, which could cut the travel time to Mars to only 100–120 days. Nuclear rockets are not really a new idea for NASA, but have been around for nearly 50 years. As early as 1961, Wernher von Braun partnered with the Atomic Energy Commission to develop nuclear propulsion, as he envisioned that NASA would be sending astronauts to Mars after the Apollo program (Cain July 1, 2019).

There was also another significant nuclear propulsion development program NASA called "NERVA" or the Nuclear Engine for Rocket Vehicle Application. It was tested in the deserts of Nevada at Los Alamos in the late 1960s. Several successful tests were demonstrated, but nothing was adopted and used by NASA. But engineers and NASA leaders were convinced that this was the type of propulsion technology needed for a quick, efficient, and successful journey from Earth to Mars.

Unfortunately, nuclear thermal propulsions rockets' time had not yet come. During the Nixon Administration, the Mars idea was tabled, and in 1971, President

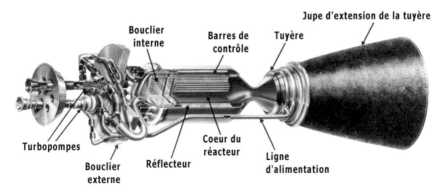

Fig. 7.2 A nuclear reactor core-powered space propulsion system. (Graphic courtesy of NASA)

Nixon chose to proceed forward with the development of the Space Shuttle. By 1973, this innovative type of propulsion was stalled and nearly forgotten (NASA May 25, 2018).

Recently, in May of 2019, NASA requested funds to begin to develop nuclear thermal propulsion rockets. Surprisingly, Congress approved the funding, and $125 million was allocated for testing and development. NASA's Artemis plan to return to the Moon would not require this type of propulsion, but NASA was beginning to look forward to future deep-space travel and be prepared for the type of short-term, long-distance journeys needed to enable humans to explore the solar system safely (Wall August 8, 2019) (see Fig. 7.2).

Another incredible benefit of nuclear rockets is that launch windows would not have to be every 26 months. Even when Earth and Mars are not aligned, the spacecraft could launch and not worry with delays. This quick method of travel would also allow astronauts to stay on the Mars surface for shorter amounts of time, and they would not need to wait for the perfect launch window to return back to Earth. Thermal nuclear rocket propulsion is a key factor for a successful journey to Mars and back.

References

Cain, F. Earth to Mars in 100 Days: The power of nuclear rockets. *Universe Today*. July 2019. https://phys.org/news/2019-07-earth-mars-days-power-nuclear.html. Last accessed 3 Feb 2020.

Foust, J. Momentum grows for nuclear thermal propulsion. *SpaceNews*. May 22, 2019.

Hadhazy, A. (December 22, 2014). *A new way to reach mars safely, anytime and on the cheap*. https://www.scientificamerican.com/article/a-new-way-to-reach-mars-safely-anytime-and-on-the-cheap/. Last accessed 9 Mar 2020.

NASA. (June 7, 2012). *Voyages: Charting the course for sustainable human space exploration*.

NASA. (May 29, 2014). *Pioneering space: NASA's next steps on the path to Mars*.

NASA. (May 25, 2018). *Nuclear thermal propulsion: game changing technology for deep space exploration.* https://www.nasa.gov/directorates/spacetech/game_changing_development/Nuclear_Thermal_Propulsion_Deep_Space_Exploration. January 31, 2020. Last accessed 23 Nov 2019.

National Academy of Sciences. (June 2014). *Pathways to exploration: Rationales and approaches for a U.S. program of human space exploration.*

Sommerer, J. C. (February 3, 2016). *Hearing of the house committee on science, space, and technology subcommittee on space.*

Wall, M. (August 8, 2019). *Nuclear propulsion could be 'Game Changer' for space exploration, NASA chief says.* https://www.space.com/nuclear-propulsion-future-spacecraft-nasa-chief.html. Last accessed 3 Feb 2020.

Chapter 8
An Economic Perspective

An analysis of contemporary reports and my interviews indicate that there is a significant economic advantage to a Moon mission versus a Mars mission. First, the estimated budget of a round trip to the Moon is substantially less than a trip to Mars. Second, a mission to the Moon is more in line with the current budgetary abilities of the US government and NASA's international partners. Also, the Moon potentially has rich raw resources that might counter-balance the expense by bringing those resources to Earth.

There is little written information on the subject of costs of going to these destinations and creating permanent habitats on the Moon and Mars. Several experts were queried, and yet no one was willing to provide written reports with regard to estimated cost analysis or a comparative analysis of the Moon versus Mars. One of the biggest challenges is the indefinite and uncertain estimates projected for a round trip journey to Mars. There are some estimates as low as $80 billion at the low end and upward of $1.5 trillion on the extreme higher level. These conflicting figures breed distrust and unwillingness among congressional leaders to commit American taxpayers' dollars to such an ill-defined mission. At a House Committee hearing on Human Space Exploration in 2016, Mr. Tom Young, former Director of Goddard Space Flight Center, stated:

> It is hard to sell a plan until you have a plan. That is kind of "step one" of the process, in my view. My other comment is that it is just not any plan; it must be a plan that people, both pro and con, can recognize as credible— and that the ingredients of the plan [truly] exist. I think there is a reasonable probability that in the next two decades we will spend $180 billion on human space exploration, which is not a bad down payment. In my view that needs to be a critical part of the plan, but I do think it will have to be augmented. (Young February 3, 2016)

Attending the same hearing as an expert witness was Dr. John C. Sommerer, former Director of Johns Hopkins Space Department Research Center, who also executed NASA's MESSENGER mission to Mercury and the New Horizons mission to Pluto. Rep. Ed Perlmutter (D-CO) voiced his concerns over the vague and confusing budget analyses of proposed Mars' missions and questioned Dr. Sommerer:

A. Reneau, *Moon First and Mars Second*, SpringerBriefs in Space
Development, https://doi.org/10.1007/978-3-030-54230-6_8

So, Dr. Sommerer, you said that according to your research and the panel's investigation
this was twenty to forty years and at least half a trillion dollars. How did you come up with
that?

Sommerer replied:

I do not want to say that it is a half a trillion. It is on the order for half a trillion, but maybe
we will get by with 180 billion. (Congressman Perlmutter and Sommerer February 3, 2016)

In addition, the public has an exaggerated perception of the amount of
taxpayer money spent by NASA. The average American thinks that the NASA budget accounts for 2.5 to 5 percent of the entire federal budget. In truth, it is hovers
around 0.5 percent (Mars Generation Survey March 7, 2013).

Whatever the costs may be, NASA needs to do a better job of communicating
financial facts in order to rally public support for future human space missions.
Sommerer continued to testify concerning limited budgets and the disoriented
vision of the future of human space exploration:

To be explicit and to set the scale of the problem, the Technical Panel, aided by independent
cost estimation contractors, and using and innovative process that respected the importance
of development risks based on technical challenges, capability gaps, regulatory challenges,
and programmatic factors, and the need to retain a reasonable operational tempo, concluded
that the first crewed Mars landing might be possibly 20–40 years from now, after a cumulative expenditure of on the order of half a trillion dollars. The actual time frame and cost will
depend greatly on the pathway chosen to achieve the goal of going to Mars and candidly,
the fastest and least expensive pathway that we examined comes with enormous risks to
both the success for the missions and lives of the astronauts conducting them.

NASA's current plans have serious deficiencies. To quote the Technical Panel's
final briefing to the entire NRC (National Research Council) Committee in 2014:
"In the current fiscal environment, there are no good pathways to Mars."

The Aerospace Safety Advisory Panel (ASAP) was founded by Congress in
1968, after the Apollo 1 fire in 1967, which claimed the lives of three American
astronauts. This panel is tasked with advising on safety protocols and giving recom-
mendations to NASA leadership. ASAP holds quarterly public meeting and conducts fact-finding operations while visiting NASA centers and identifying potentially
dangerous and hazardous situations. In its 2015 annual report, the Panel expressed
concern that NASA lacked detailed plans in the areas of technology, vehicle design,
and the agency's budget and articulated some reservations about NASA's ability to
carry out a successful manned mission to the red planet. And what were the panel's
primary reasons for existence? The answer given was NASA's inability to provide
adequate details in two areas: technology and budget (Aerospace Safety Advisory
Panel NASA, January 23, 2020).

When NASA's leadership was asked to comment on this report, they said it was
too early to create a detailed plan. They said they were reluctant to design spacecraft
and technologies needed for a mission to Mars, citing that they expect technologies
to advance greatly in the next two decades. There was also deep concern about a
decision by the next presidential administrations to eliminate what NASA may plan
now. But the ASAP panel believes if a mission—any mission—is well designed,

Fig. 8.1 NASA's Vehicle Assembly Building at the Kennedy Space Center. (Graphic courtesy of NASA)

with supporting facts and figures, NASA will receive support from the next president (Marie Doctor January 19, 2016).

At Kennedy Space Center in April 2010, President Obama reiterated his commitment to a manned Mars' mission (see Fig. 8.1):

> By 2025, we expect new spacecraft designed for long journeys to allow us to begin the first-ever crewed missions beyond the Moon into deep space. We'll start by sending astronauts to an asteroid for the first time in history. By the mid-2030s, I believe we can send humans to orbit Mars and return them safely to Earth. And a landing on Mars will follow. And I expect to be around to see it. (President Barack Obama April 15, 2010)

Regrettably, since this announcement, NASA became a casualty of budget cuts that would have long-lasting impacts on spacecraft designed for long-distance deep-space missions. As these cuts found their way into Mars manned missions, it began to determine when humans might navigate their way to this challenging destination. A manned mission to the red planet requires an enormous amount of research, development, and financial investment. US space policy between 2010 and 2016 did not appear to have the political or fiscal will to commit to such an ambitious goal. Ayanna Howard, Chair of the Robotics Doctoral Program at Georgia Institute of Technology, stated:

> Unfortunately, development is closely tied to budget . . . . If sufficient funding is made available, then scientists and engineers should be able to develop and integrate the required EDL

components necessary for a human Mars mission within the next thirty years. If not enough resources are allocated, this timeline might not be feasible. (Herath April 18, 2011)

At a minimum, the US spaceflight program budget needs to grow at the rate of inflation. Equally important, a plan needs to be developed that demonstrates a reasonable timeline that is immune from partisan politics. The sustainable path to deep-space, manned exploration depends on a strategy where stakeholders from government, industry, international partners, and the public are vested in the program's success. The power of partnership will maintain ambitious human exploration plans through its ups and downs as proven by the collaboration of many nations and private companies invested in the success of the International Space Station. A mission to the Moon, Mars, or beyond should not be any different. It is my belief that the United States and NASA should lead the charge.

The following statement by the Technical Panel at the hearing of the House Committee on Science, Space, and Technology summed up the current situation rather well:

> I would like to conclude with some of my own views. I understand that there is bipartisan support for a 'go as we pay' approach to human spaceflight. But just as it is not feasible to take a cross-country trip on a child's allowance, because of threshold costs, we may well never be able to get to Mars at our current expenditure level. It might be better to stop talking about Mars if there is no appetite in Congress and the Administration for higher human spaceflight budgets; and more disciplined execution by NASA. At a minimum, we should agree on a pathway that is satisfying to the public, even if it does not lead to Mars in the foreseeable future. A pathway that includes the surface of the Moon is one obvious possibility. (Sommerer February 3, 2016)

Indeed, there are budget constraints in our current political cycle, but we have many advantages we did not have in the 1960s during the Apollo era. We have 50 years of developing new space technologies, thanks to companies like SpaceX and Blue Origin. Through the miraculous reality of reusable rockets, miniature electronics, and artificial intelligence, we can go much swifter with much less expense. With this new and exciting, fast-paced private space sector, it truly is possible to put astronauts on the Moon by 2024—and it can be done efficiently, on schedule, and under budget.

References

Aerospace Safety Advisory Panel, NASA. (January 23, 2020). http://oiir.hq.nasa.gov/asap/. Last accessed 13 Feb 2020.

Herath, A. K. "Why is it So Hard to Travel to Mars?" *Astrobiology Magazine.* April 18, 2011. http://m.space.com/11417-Mars-mission-space-travel-challenges.html. Last accessed 15 Oct 2019.

Doctor, Marie R. Safety panel doubts NASA's capability for 2030 manned Mars mission. *Tech Times.* January 19, 2016. Last accessed 6 Feb 2018.

Mars Generation Survey. (March 7, 2013). http://www.exploremars.org/wp-content/uploads/2013/03/Mars-Generation-Survey-full-report-March-7-2013.pdf. Last accessed 12 June 2019.

Obama, B. (April 15, 2010). *Space exploration in the 21st century.* http://www.nasa.gov/news/media/trans/obama_ksc_trans.html. Last accessed 6 Feb 2018.

Perlmutter, E. D., & Sommerer, J. (February 3, 2016). Hearing of the House Committee on Science, Space, and Technology Subcommittee on Space. *Charting a course: expert perspectives on NASA'S human exploration proposals.*

Sommerer, J. C. (February 3, 2016). Hearing of the House Committee on Science, Space, and Technology Subcommittee on Space.

Young, T. (February 3, 2016). Hearing of the House Committee on Science, Space, and Technology Subcommittee on Space.

Chapter 9
Private Lunar Initiatives

The Moon was initially viewed as a trophy for a political, military, and technological contest between Russia and the United States. But with the recent discovery of billions of tons of water ice in the shadowed craters of the north and south poles of the Moon, there is a new "Gold Rush" or "Moon Rush" for these valuable resources. Space exploration and tourism are now entering a new chapter driven by business plans and personal profits. Private American space companies are eyeing the Moon as a fueling station for their ambitious private flights to Mars, as well as fuel and oxygen for humans living and working on the lunar surface.

As the global space economy tops out at nearly 400 billion in 2020, the next decade could easily approach the trillion-dollar mark. The future for commercial space companies has become bright and optimistic, and Kennedy Space Center and the Cape are once again alive with enthusiasm, optimism, and innovation.

Space tourism will become a reality with the first suborbital excursions possibly occurring by 2021. Many potential customers have already purchased their tickets in advance to the tune of $250,000 each. Soon, SpaceX and Boeing (in partnership with NASA) will launch American astronauts from American soil. Both of these commercial companies, as well as Northrop Grumman Innovation Systems, currently fly resupply missions to the International Space Station. Roscosmos, a Russian space agency, even offers spaceflights for private citizens to the ISS. Increasingly, government space agencies are contracting with private space corporations for technology and transportation.

Russia and China continue to challenge the United States for global leadership in space, and more than 70 other countries are progressing rapidly forward. But this new space race is no longer just between governments; it's becoming a competition between private companies also, to see who will get to the Moon first—as well as asteroids. Mining companies here on our home planet may face challenges as these companies seek to mine precious metals on other planetary bodies, the Moon, and near-Earth asteroids. Ultimately Earth will be depleted of its natural resources, and

A. Reneau, *Moon First and Mars Second*, SpringerBriefs in Space Development, https://doi.org/10.1007/978-3-030-54230-6_9

commercial companies will continue to innovate and create solutions to guarantee the future of our planet.

These nimble entrepreneurial organizations are finding ways to travel to space more efficiently and at far less cost than our bureaucratic, slow-moving government agencies. Many countries are relaxing regulations and are creating supportive and enabling environments which encourage, rather than suffocate, the private sector. And with the passage of Space Policy Directive 2, these companies have been unshackled. The world is experiencing a unique time in history with the perfect combination of available funding, political will, decreasing costs, and cutting-edge technology.

For several decades there have been numerous well-established companies who have partnered with government and the military complex to push forward the political goals of multiple nations. These include Airbus, Arianespace, Boeing, Lockheed Martin, and Northrop Grumman to name a few. But in the last decade, there were some "new kids on the block."

SpaceX

In addition to NASA and its international partners striving for a quick return to the Moon, multiple private companies are on the same quest, albeit for different reasons. The majority of these companies are backed by notable billionaires, each with a unique and ambitious vision. The most well-known private space company is SpaceX, which was founded in 2002 by Elon Musk in Hawthorne, California. Musk's main goal has always been the colonization of Mars (Hofford March 13, 2019).

Although this goal may sound like a science fiction movie, Mr. Musk always comes through to the finish line with his bold declarations—even though there is always some delay. A charismatic performer combined with a brilliant engineering team, Elon launched his bright cherry Tesla Roadster as a test payload aboard the Falcon Heavy rocket in February 2018. (See Fig. 9.1.)

I was present at Kennedy Space Center when he flawlessly placed this sports car and its front seat driver, Starman (the astronaut mannequin), into Mars orbit. SpaceX then successfully landed the two Falcon Heavy rocket boosters on pads—simultaneously—at Kennedy Space Center right before our very eyes. Now SpaceX is in the process of developing the StarShip, the largest rocket to ever be built. Additionally, this rocket will have other missions such as transporting humans on tourism missions around the Moon and one day to Mars:

> When Henry Ford made cheap, reliable cars people said, 'Nah, what's wrong with the horse?' That was a huge bet he made, and it worked! (Musk February 4, 2017)

Fig. 9.1 Falcon Heavy launcher. (Graphic courtesy of SpaceX)

Blue Origin

The second company with a little less flare and desire for public attention is Blue Origin, established by Amazon founder and billionaire, Jeff Bezos. A graduate of Princeton University, Bezos has always had a passion for space and rockets. His vision is to have millions of people living and working in space, and most recently, he revealed his Blue Moon Lander which has been in development for several years. Mr. Bezos' goals align more with NASA to enabling humans to land, live, and work on the Moon. Blue Origin would like to provide Earthlings the joy of space tourism and continue to drive down the cost for space travel. "Personally, I would love to go to space," Bezos says. "But it's not the thing that is most important to me. I believe that we are sitting on the edge of a golden age of space." Just as he used the Internet to save consumers a substantial amount of money, he also believes his theory of reusability will make space accessible for middle class people. He is sure that once travel to space is affordable and safe, a whole new industry will be created that have not even entered our minds yet. "I don't want to live in a civilization of stasis," Bezos says. "I want to live in a civilization of invention, and growth, and incredible new things. And I am very confident it's the only way. We have to go to space" (Fishman December 2016).

Fig. 9.2 Blue Origin launcher headed aloft. (Graphic courtesy of Blue Origin)

Moon Express (Fig. 9.3)

Another privately held company co-founded by Bob Richards, Naveen Jain, and Barney Pell in 2010 is named Moon Express. This private enterprise originally had the goal of winning the Google Lunar X Prize, with the hopes of capturing the $30 million prize money. Although they were not able to complete the challenges prescribed for this mission before the deadline and collect the cash prize, Moon Express still has a plan to launch commercial payloads to the lunar surface in the early 2020s and was awarded a NASA contract under the Commercial Lunar Payload Services agreement. This program allows Moon Express and other companies to bid on carrying science payloads to the Moon. In addition, the company has goals to mine resources on the Moon as well as a robotic mission named Lunar Outpost MX-3 to scout for water ice at the Moon's South Pole. Ultimately Moon Express would like to conduct a sample-return mission of lunar resources:

> The challenge of securing humanity's future through our expansion into space and ultimately becoming a multi-world species is both necessary and noble. But it's not about boldly going; it's about boldly staying. To go to space to stay, it has to pay. (Richards September 7, 2017)

For over half a century, the Moon has been a frontier for governments only. But that is now changing. There is a budding private space sector known as Space 2.0 with a strong interest in finding new applications for new space technology. There are now many new and ambitious business and economic goals in space. One of

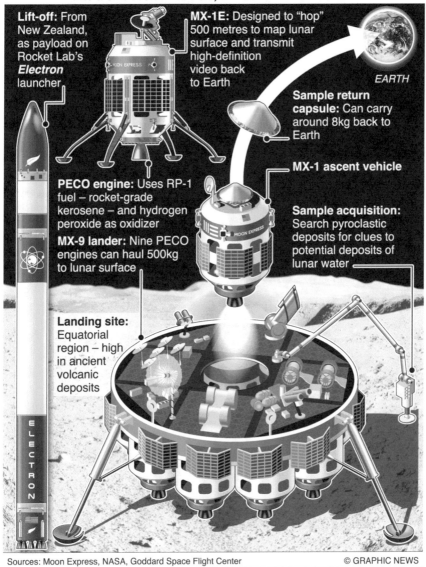

Moon soon to be open for business

Moon Express hopes to become the first private spaceflight company to send a robotic lander to the moon, scoop up some moon dust, load it into a small return vehicle, and send it back to Earth

Lift-off: From New Zealand, as payload on Rocket Lab's *Electron* launcher

MX-1E: Designed to "hop" 500 metres to map lunar surface and transmit high-definition video back to Earth

EARTH

Sample return capsule: Can carry around 8kg back to Earth

MX-1 ascent vehicle

PECO engine: Uses RP-1 fuel – rocket-grade kerosene – and hydrogen peroxide as oxidizer

MX-9 lander: Nine PECO engines can haul 500kg to lunar surface

Sample acquisition: Search pyroclastic deposits for clues to potential deposits of lunar water

Landing site: Equatorial region – high in ancient volcanic deposits

Sources: Moon Express, NASA, Goddard Space Flight Center © GRAPHIC NEWS

Fig. 9.3 Moon Express is seeking regulatory permission for lunar flight. (Graphic courtesy of Moon Express)

these interests for new space companies includes lunar exploration and development. Some of these Space 2.0 ventures are beginning to consider the Moon a "new continent" which holds valuable resources that could bring benefit back to the inhabitants of our delicate planet. Their perspective is that opportunities and boundless resources await humanity. These new entities focused on space development see the impending exploration of the Moon as a means to harness new and clean energy for human beings on planet Earth and much more.

References

Fishman, C. Is Jeff Bezos' blue origin the future of space exploration? *Smithsonian Magazine*. December 2016.

Hofford, C. (May 13, 2019). *Top 3 biggest private space companies*. https://www.earth.com/earth-pedia-articles/top-3-biggest-private-space-companies/. Last accessed 16 Dec 2019.

Musk, E. via Twitter. February 4, 2017.

Richards, B. (Thursday September 7, 2017). Testimony before the US House of Representatives Committee on Science, Space and Technology Subcommittee Space Hearing on Private Sector Lunar Exploration.

Chapter 10
An International Perspective

The outcomes of America's space-related accomplishments are not achievements of the United States alone. NASA's success in space has always been part of an internationally shared endeavor—including the Apollo program. Former Apollo 17 astronaut Harrison Schmitt made this observation:

> A lot of people don't know that there has always been international cooperation. NASA has always used other countries—as they did for the Gemini and Apollo programs. From a geopolitical perspective we have to cooperate in the future. We should offer opportunities for other nations to participate. But I think if you try to manage future missions internationally, it is doomed to failure. You have to have a designated leader. (Schmitt July 28, 2015)

The ISS is a tribute to the shared cooperation and sacrifices of many countries which is a crowning triumph in the arena of worldwide collaboration. With over 18 countries involved, the ISS just celebrated 19 consecutive years in low Earth orbit. Together, significant scientific breakthroughs have been achieved that have transformed how human beings live on Earth, as well as groundbreaking research on the effects of microgravity on the human body over long periods of time.

As the technological revolution accelerates, many countries are partnering to operate a variety of technologies, including global navigation systems. Space-faring nations are on the rise, and certain countries no longer hold a monopoly on technology in space. We must find a way to continue the pattern set by the ISS in order to guarantee that competition does not overshadow international cooperation. Deputy Secretary of State William Burns noted:

> The International Space Station remains a leading space platform for global research and development. The Station is the foundation for future human exploration to an asteroid, the Moon, and ultimately to Mars. It is a lasting testament to how much more we can accomplish together than we can on our own. (Burns January 9, 2014)

A. Reneau, *Moon First and Mars Second*, SpringerBriefs in Space
Development, https://doi.org/10.1007/978-3-030-54230-6_10

One group focused on international cooperation to maintain openness and inclusiveness is the International Space Exploration Coordination Group (ISECG), a consortium of 15 space agencies. The following space agencies are ISECG members (in alphabetical order): ASI (Italy), CNES (France), CNSA (China), CSA (Canada), CSIRO (Australia), DLR (Germany), ESA (European Space Agency), ISRO (India), JAXA (Japan), KARI (Republic of Korea), NASA (United States), NSAU (Ukraine), Roscosmos (Russia), UAE Space Agency (United Arab Emirates), and UKSA (United Kingdom).

ISECG was created on the platform of a shared vision of coordinated human and robotic space exploration focused on solar system destinations where humans may one day live and work. ISECG is a voluntary, non-binding international coordination effort through which the individual agencies may exchange information regarding their interest, plans, and activities in space exploration. This international organization works together to focus on strengthening both individual exploration programs and the collective effort (Mahoney February 2, 2018).

One former leader of ISECG, Kathy Laurini, said:

> I worked for ISECG and led NASA's engagement in the ISECG, which is a group of national space agencies that have common desire for human space exploration. We try to create the foundations and partnerships with NASA as a leader in space exploration. This keeps our international partners informed on what NASA is doing. The other space agencies bring a lot of good ideas to the table and they have a desire to be part of a bigger effort as we go forward. This consultation and consensus we do builds collaboration to help push forward our common goals. (Laurini July 28, 2015)

The Global Exploration Roadmap produced by ISECG in 2019 is a well-laid-out plan that conveys the platform and strategy for coordinating both robotic and human exploration throughout the solar system. It reflects international preparedness for cooperative space missions to the Moon, asteroids, Mars, and beyond. Space agencies worldwide agree that human space exploration will be more successful if it includes many nations because there are so many obstacles that stand in the way of accomplishing these missions. In addition, they agree that by pursuing these goals, the benefit to people on Earth is quite extraordinary—intellectually, culturally, socially, and economically. ISECG meetings have generated innovative ideas, as well as thoughts for problem-solving, and in turn have strengthened relationships among the member space agencies. These partnerships will be required for a sustainable human presence in space. Yang Liwei, China's first taikonaut, who flew the Shenzhou mission in October 2003, stated: "I think the development of space endeavors is not for one nation or one country. I myself, as an astronaut, believe that the multinational, the international cooperation, is the future triumph of the development of space industry" (Moskowitz April 29, 2010).

Brief Examination of Lunar Exploration Programs from Around the World

The prime focus of this book on lunar exploration and development, plus follow-on efforts to go to Mars, has been primarily sought to examine US governmental space activities and especially those of NASA. This is because there is a great deal of activity in this area at this time with Artemis, Lunar Gateway, and other related efforts. Furthermore, US national space policy and regulation is where the expertise of the author of this book predominantly lies. This is not to discount or ignore the very considerable international efforts that have been ongoing to send exploratory missions to the Moon. In addition, there is also a growing enterprise of strictly commercial initiatives to explore and even create permanent facilities on the Moon. This chapter will seek to relate some of the important ongoing and growing international efforts to explore the Moon. We will, in particular, examine current lunar programs of China, Europe, India, Israel, Japan, Russia, and the UAE, which all have plans in place to pursue lunar and deep-space activities. In some cases, these national space programs are in the form of joint international space programs that will be so noted through the structure of this chapter, as arranged in a country-by-country basis. This is not to say that other national space programs involving the Moon and deep-space initiatives do not exist, but the following examination seeks to highlight those that are presently at the top of these international efforts in terms of funding and current level of effort.

China

The scope of Chinese space programs continues to expand exponentially. No other national space program has grown more rapidly in terms of level of technical, operational, and financial activity than that of China. President Xi Jinping praised the outstanding excellence of the Chinese space program and especially the Chang'e 4 lunar lander and Yutu-2 lunar rover that successfully landed and explored the far side of the Moon in January 2019. He even suggested that the Chinese Communist Party, the national armed forces, and all ethnic Chinese people should emulate as a "model" the excellent accomplishments of the Chinese space program.

If there is a true rivalry in space programs around the world today, it seems likely that most experts would identify the national space programs of the United States and China as being engaged in competitive initiatives. It may well be true that in another decade this may even be true with regard not only to national space programs but to commercial space initiatives as well. Indeed, private Chinese space initiatives are now rapidly multiplying, along with the official national space programs of the Chinese government (Campbell July 17, 2019).

The follow-on to the aspiring Chang'e lunar programs is now a specifically announced objective of creating a permanent scientific research station near the

Moon's South Pole. There is a belief among space policy analysts that these expanding space-related goals are central to Chinese basic national economic expansion goals. The recent *Times Magazine* article on the rise of the Chinese space program stated the context in this way: "The speed at which China is surpassing each technological hurdle spotlights how the Beijing government views space as vital for boosting the economy and promoting high-end industry and spin-off technologies" (Campbell July 17, 2019).

The Chang'e 1 lunar lander in 2007, the Chang'e 2 in 2010, the Chang'e 3 in 2013, and the Chang'e 4 in 2019 are clear evidence of China's ambitions to pursue an ongoing exploration of the Moon. This is to be followed by the explicitly announced goal of creating a robotic research station near the water reserves close to the lunar South Pole as the next step of exploration and further development of the Moon. The Chang'e 5 mission is scheduled to land in 2020 north of the lunar equator on the wide huge basaltic plain that is known as Oceanus Procellarum and provide a sample return to Earth and is also slated to provide a sample return from the Moon's South Pole. Chang'e 7 will include a rover which will explore the lunar South Pole region with the likely next step being the creation a robotic scientific research station in this area. There are indications that there may be a joint Chinese and Russian effort in lunar exploration with regard to these activities.

There are now at least three space initiatives heading in the direction of the lunar South Pole. NASA has indicated that it plans to send two astronauts to land near this area as early as 2024. The European Space Agency has expressed a plan to create a "Moon Village" in this region as well. As noted above, there may be Russian cooperation in the Chinese lunar program.

All the activity in the southern polar region of the Moon leads to concerns that the international space programs should be in collaboration so that landing and take-off of vehicles from these research stations do not contaminate the other sites. Further, there are other non-governmental initiatives that have been variously identified by commercial organizations such as Moon Express, Shackleton Energy Company, Golden Spike, and others that have expressed possible lunar enterprises as well (Wall July 18, 2019).

Europe

The other significant player in lunar exploration is Europe. These lunar initiatives are coming primarily through the European Space Agency, but also involve various other European national space agencies. These include France, Germany, Italy, Spain, Switzerland, and the United Kingdom, even though there are other European space agencies that may provide some level of support.

At the International Astronautical Congress (IAC) held in Washington, D.C., in October of 2019, NASA Administrator James Bridenstine indicated that 26 nations, with most being from Europe, had expressed interest in the lunar exploration Artemis program and that specific agreements to solidify the nature of that cooperation would be negotiated quickly. In this presentation Bridenstine indicated that one approach might be to expand the International Space Station's Intergovernmental Agreement (IGA). Signed in 1998, this includes 15 nations as a starting point. He also explained that private entities were anticipated to be a part of the Artemis program as well as nation states. NASA Administrator Bridenstine declared: "The goal is to have many different nations living and working on the Moon at the same time with a coalition…" (Howell October 25, 2019).

Although this Artemis coalition would likely include Japan, Canada, and other space-faring countries, it is clearly expected that the "center of gravity" in the international coalition would be provided by European countries, especially in the form of the European Space Agency (ESA) and other leading European national space agencies.

Johann-Dietrich Woerner, Director General of ESA, has ambitions that go beyond a simple research station on the Moon. Since the IAC in October 2015, he has lobbied for a combined global initiative to create an international "Moon Village" that could accommodate astronauts from all over the world which could live together in a facility such as the one pictured in Fig. 10.1.

Director General Woerner, speaking at the Space Tech Europe Conference in October 2017, explained his vision as follows: "I would like to see enough natural intelligence to have a different scheme for cooperation than just competition….. Space is above all Earthly borders, and if we don't cooperate, how can we expect other areas to cooperate?" Woerner's speech was seen as a counter-balance to some of the calls from US leadership for further space competition. He explained that his vision was global cooperation to clean up space debris and create a "Moon Village" that would be open to China, India, and others beyond the usual types of space

Fig. 10.1 Artist conception of a "Moon Village" which could be an international settlement on the Moon. (Graphic courtesy of ESA)

partnerships that have characterized the International Space Station (ISS) franchise agreements (Pultarova Oct. 24, 2017).

There are, however, more concrete steps currently underway for not only ESA but European aerospace commercial contractors to be a part of Artemis lunar exploration program. Key to this effort is the Lunar Gateway project which is a lunar orbiting space station designed to house astronauts and serve as an orbiting habitat and staging area for lunar experiments and exploration. If Gateway remains part of the Artemis program, this facility would serve as a way station, while astronauts and robotic systems construct a more permanent shelter for experimental equipment and housing facilities in the region of the Moon's South Pole. At this stage European participation significantly involves Airbus's active involvement in the design and fabrication of the service module for NASA's Orion Multi-Purpose Crew Vehicle. Orion would ride atop the Space Launch System (SLS) heavy-lift rocket that supports the necessary launches to lunar orbit to deploy the Lunar Gateway modules and eventually create the new habitat facility on the lunar surface. Other aspects of European participation will be defined as further international agreements are signed between the United States, European space agencies, and relevant commercial contractors.

India

Just as China has shown significant new technological, scientific, and operational capabilities in space in the past two decades, there have been key developments of Indian space initiatives involving both the Moon and Mars in recent years. ISRO's first mission to the Moon was the Chandrayaan-1 that was a lunar orbiter. This was launched on October 22, 2008, to survey the Moon's surface using the Polar Satellite Launch Vehicle (PSLV-C11). This satellite continuously operated for some 10 months, and during this time, it made more than 3400 orbits around the Moon. It was thus able to map the lunar surface in great detail to create an accurate three-dimensional map of the surface features. The orbiter was also able to create an accurate assessment of the minerals located on the surface and allow inferences to be made of what might be below the surface as well. The Indian Space Research Organization (ISRO) that designed the spacecraft and launched it into lunar orbit indicated that its most significant achievement was the detection of both hydroxyl (OH) and water (H_2O) especially in the southern polar region of the Moon. It also detected aluminum, silicon, calcium, and magnesium. Chandrayaan-1 provided sufficient evidence that Indian scientists had previously suggested, that there are large ice deposits below the surface—particularly in craters not illuminated by the sun (Results from Chandrayaan-1 Million 2020).

This mission also supported operational experience to support communications between the orbiter and the Indian Deep Space Network (IDSN). This network includes both an 18 m and 32 m antennas that are linked to the Indian Space Science

Data Centre (ISSDC) at Byalalu. This ISSDC represents the primary data and digital analysis center for Indian space science and exploration missions. This center is the one that is used for the follow-on of Chandrayaan-2 and Chandrayaan-3.

The Chandrayaan-2 mission was successfully placed into lunar orbit in 2019. This mission included an orbiter space craft, a lander system, and a rover. Unfortunately, due to command errors related to the lander, it resulted in a hard landing on September 2019, which ended this part of the mission. However, the orbiter continued to operate and collected data that was comparable to that collected by Chandrayaan-1. The lander's cameras also captured useful data of the area of its descent prior to the unsuccessful landing.

As of January 1, 2020, the Indian government announced it is now planning for a Chandrayaan-3 mission. This project will also consist of an orbiter, a lunar rover, and a stationary lander that is in many ways very similar in design to the Chandrayaan-2 spacecraft. India's official determination to try again was announced by Kailasavadivoo Sivan, the Chairman of the Indian Space Research Organization (ISRO), at a news conference on New Year's Day, 2020. This decision was made less than 3 months after the earlier Chandrayaan-2 lander's hard landing on September 2019 (Wall Jan. 2, 2020).

India's deep-space missions have also included a very ambitious mission to Mars. The Mars Orbiter Mission, known as Mangalyaan, was launched on November 5, 2013, and was placed into Mars's orbit on September 24, 2014. This space probe was a project of the Indian Space Research Organization (ISRO) and represented the fourth nation to achieve this objective after the United States, Russia, and Europe. Mangalyaan was notable in that the cost of the deep-space probe was far less than that of other countries and was achieved on the first attempt to do so.

The Mars Orbiter Mission (known as MOM) is currently mapping the surface of the distant red planet. The scientific data from the mission will provide additional mapping information of Mars and also test technologies and operational systems that will be needed for future Indian space probes (Indian Mars Orbiter Mission 2020).

Israel

Israel is one of the new space-faring nations. Nevertheless, this small country has long expressed strong interest in lunar research and development. Israel, through governmental initiatives and especially via private space companies, has sought to advance lunar research as well as future efforts to undertake commercial space development projects on the Moon. SpaceIL is a not-for-profit Israeli space organization that has been funded by private donors. It was established in 2011 for the purpose of competing for the Google Lunar X Prize contest. Its stated purpose was landing a spacecraft on the Moon and additionally to provide video coverage of the original Apollo landing site. This Google Lunar X Prize competition offered a potential winner a $30 million prize and led to over a dozen competitors formally qualified to contend for the prize. This incentive led to the development of the Beresheet spacecraft by SpaceIL. Eventually, the competition officially ended at the end of 2018; the SpaceX organization announced that it would contribute $1 million to SpaceIL to continue its work in spite of the Beresheet spacecraft crashing into the Moon's surface rather than making a soft landing. This spacecraft was launched on a Falcon 9 launch vehicle that also launched other satellites and then was spiraled out to a point where it could be injected into lunar orbit (See Fig. 10.2).

Beresheet was the first Israeli spacecraft to travel beyond Earth's orbit. It was based on private development funding, since this was a Google SpaceX competition requirement. This was thus the first privately funded space vehicle to land on the Moon, even though it broke apart when an engine malfunctioned and it crash-landed on the Moon's surface on April 11, 2019. The budget for this program was reportedly some $100 million dollars which would make it the most modestly funded lunar project in history (Crane April 11, 2019).

In addition to the SpaceIL Beresheet project, there is a quite active Israeli space project backed by the Israeli Space Agency (ISA) which operates the Shavit launch vehicle that is designed primarily to place satellites into Earth orbit. ISA provided some support to the Beresheet 1 and has also committed to provide on the order of $5.6 million to the Beresheet 2 lander that is currently being developed by SpaceIL. NASA has also supported the project in providing laser communications technology that is being used in this lunar exploration program. The date for the second Beresheet launch has not yet been firmly established (Oster May 08, 2019).

Fig. 10.2 A "selfie" image from the Beresheet lunar lander sent prior to its crash landing on the Moon. (Graphic courtesy of SpaceIL)

Japan

Japan has closely aligned its space research activities with the United States and NASA programs, but it has also undertaken its own deep-space research activities. One of its key space research projects that explored the Moon and its make-up was the Selene (or Kaguya) project which was carried out by the Japan Aerospace Exploration Agency (JAXA).

This satellite was launched by the Japanese on September 14, 2007, and was successfully launched from the Japanese Tanegashima Space Center (TNSC). The project sought new scientific data about the lunar origin, the composition of the Moon's surface, and its evolution over time. The mission consisted of a main orbiting satellite that began operation at about 100 km altitude and then lowered to a 50 km orbit in February 2009 and then to an even lower orbit and then deteriorated to 30 km and 10 km and eventually crashed into the Moon on June 10, 2009. In addition, there were two small satellites known as Relay Satellite and the VRAD Satellite that were deployed in a polar orbit around the Moon to aid with visibility to the Earth and expanded coverage. The orbiters provided a range of scientific and telecommunications relay services and instruments to allow extended scientific investigation of the Moon and then relay data back to Earth more efficiently (Kaguya (Selene) January 20, 2020) (see Fig. 10.3).

The Selene mission, in addition to providing imaging of the Moon, also investigates energetic particles, electromagnetic field, and plasma around the Moon. The measurements of the lunar environments are highly valuable scientifically and also provide important information for the future human activity on the lunar surface. This research provided critical research about solar wind and radiation hazards and

Fig. 10.3 Selene (Kaguya) image of the lunar surface from low altitude. (Graphic courtesy of the Japanese Aerospace Exploration Agency (JAXA))

is extremely useful for understanding the radiological dangers to astronauts on the Moon concerning life-threatening solar storms (Kaguya (Selene) January 20, 2020).

Japan via JAXA and its Institute of Space and Astronautical Sciences (ISAS) now has plans for a lunar lander to be launched to the Moon's surface in the 2020s. The first of these will be called Selene 2, and it will involve a lander project. Selene 3 would involve not only a lander but also a capability that would be designed to provide a possible like a 100 gram sample return. Finally, there is a less defined Selene X that would have a number of possible optional missions.

These options are broadly described as:

"Option-1: Technology demonstration for building outposts such as the excavation for construction of infrastructures.
Option-2: Logistics capability demonstration for building common landers for both transportation and JAXA's own robotic missions.
Option-3: Highly sophisticated insitu robotic lander, or returning samples of the surface soil to the Earth, including the development of high-speed reentry capsules." (Kubota et al. 2020)

Japan is considering and planning many more space exploration projects. There is a potential Mars mission that would include a lander that Japan has had under study for a number of years, and this might include a cooperative project with NASA. Furthermore, Japan and JAXA might well sign on to support to the NASA Artemis Project.

All of the lunar space projects from Japan are not restricted to those of the Japanese government and JAXA. There is also the commercial effort of the iSpace company of Japan. iSpace is currently planning for a lunar space probe known as the Mission 1 or M-1 project which consists of a lunar lander to be launched in 2021. This would be followed by Mission-2 or the M-2 space probe.

At the Washington, D.C., International Astronautical Conference held in October, 2019, Takeshi Hakamada, the chief executive of iSpace company of Japan, described the progress that is being made on the fabrication and testing of the Hakuto-R lunar landers that his company hopes to land on the Moon starting with the M-1 lander scheduled for 2021 landing and the M-2 scheduled for a 2023 landing. The M-2 has additional payloads and is intended to carry a lunar rover. These are scheduled to be launched on Falcon 9 launch vehicles. This project was born out of the Google Lunar X Prize competition and was first known as Team Hakuto, which evolved into the iSpace company (Frost Oct. 24, 2019).

Russia

The Soviet space program during the years of the USSR had a strong interest in competing with the United States in sending missions to the Moon. The unmanned Soviet Luna 2 landed on the Moon on September 13, 1959, and was the first to achieve this feat. Cosmonaut Leonov had been selected to go to the Moon, but this Soviet manned lunar landing was suspended when the US Apollo program achieved a series of successes between 1969 and 1972. Soviet government for years publicly denied participating in such a lunar competition, but under the Glasnost in this period, it was revealed that there had been two secret Soviet Union programs in the 1960s through the early 1970s.

These projects included a crewed lunar flyby missions known as Zond. This involved the Soyuz 7K-L1 spacecraft launched by the Proton-K rocket. There was an even more aggressive manned lunar landing program. This involved using Soyuz 7K-LOK capsule and the LK Lander. There were several failures with the N-1 rocket, and eventually both Soviet programs were ended after the US Apollo landing successes. The Zond program based on the Proton launcher was canceled in 1970. In 1974 the N1, and the related L3, ended in 1974 and was officially canceled in 1976. The Luna program proved more successful, and in 1966 the USSR Luna 9 was able to make a soft landing on the Moon and provided images from the surface. This continued with the Luna 10 through the Luna 13 missions in the next few years. There were, however, no crewed landings, due to the problems with the Zond initiative and especially with the N-1 rocket that had several major failures.

Lunar space program historian John Logsdon and NASA historian Roger Launius have both confirmed that President Kennedy, in his first meeting with Nikita Khrushchev in 1961, suggested to the Soviet leader the idea that there might be a way to make the Moon landing program a joint US-USSR project. Khrushchev, however, responded that a nuclear missile agreement and nuclear test ban document would have to be agreed upon first. Thus, such a program never happened, even though the Apollo-Soyuz mating did subsequently occur several years later in 1975 (Knapton July 13, 2019).

After this concentrated lunar research and exploration in the 1960s and 1970s, there has been a long gap in such activities. During the Putin era, Russian focus has been on space activities related to Earth. This has led to the re-establishment of a fully operational GLONASS space navigation system and the development of new space-related missile systems and hyper-velocity weapons systems. Most recently there have been indications that there is a reawakened interest in the Moon.

Dmitry Rogozin, the current head of Russia's space agency, Roscosmos, has announced long-term Moon plans: a lunar orbiter to be launched in 2024, a sample-return mission to the Moon that might come in 2028, and possibly cosmonaut missions that would come at the end of the decade. Rogozin and Zhang Kejian, head of the Chinese space agency, have announced plans for a joint lunar research initiative with data systems and scientific facilities for lunar and deep-space research, using facilities and hardware created in both countries (Bartels September 20, 2019).

UAE

The United Arab Emirates has indicated that it plans to be a key player in the space exploration and development programs of the twenty-first century. This nation has created a national space agency and is focused on the very difficult task of sending a probe to Mars. This UAE probe named "Hope" is to be launched during 2020 and reach Mars in 2021. Its announced goal is to create a Mars settlement there in the year 2117. The UAE has also proposed the creation of a Pan-Arab Space Agency, but there has not been a great deal of progress in this direction. It has also passed legislation that would support private or public ventures to engage in space mining, either on the Moon or other celestial objects, but it is unclear as to any specific goals to engage in lunar mining in the near term.

At this time the focus by the UAE has been on exploration of Mars and not on the Moon. Yet, if it is recognized in the future that much of the lunar exploration and scientific experiments related to the Moon are, in many ways, a prelude to going to Mars, it seems likely that the UAE Space Agency will turn its attention to lunar projects to a greater degree. It might, for instance, seek to be a partner in the NASA Artemis project (Rehm October 21, 2019).

Conclusions

Certainly, there are other space programs and space agencies around the world that are interested in deep-space exploration and the acquisition of more scientific knowledge about the Moon. The above reviews are only intended to give a general feeling about the breadth and depth of international space activities related to the Moon. There may be a number of activities such as Artemis that will attract broad levels of international participation as demonstrated by NASA's announcement that 26 countries have expressed interest in this US initiative. In addition, there are other efforts to build international cooperative support for lunar exploration, scientific experimentation, and other future projects. One such effort that is spearheaded by space groups in Hawaii is known as the International Lunar Decade (ILD). The objective of this program and other activities such as those of the United Nations Committee on the Peaceful Uses of Outer Space (UNCOPUOS) is to encourage international cooperative ventures in the future exploration of the Moon and the international scientific knowledge that these activities can produce.

Clearly, a number of countries understand that there is much practical and scientific knowledge that can be acquired by seeking to create lunar scientific stations and even more ambitious projects such as the Moon Village as proposed by the European Space Agency. This information can provide a key, and even vital wealth of knowledge about human life and survival in outer space, before one takes on the truly difficult task of creating permanent colonies on the Moon. Cooperative space programs involving the exploration of the Moon can likely assist humanity toward the achievement of longer-terms goals associated with the settlement of Mars. There is much to be learned by taking baby steps first. This may apply not only to technology and human physiological science but to ways to accomplish international cooperation in space as well.

References

Bartels, M. (September 20, 2019). *Russia and China are teaming up to explore the Moon*, Space. com. https://www.space.com/russia-china-moon-exploration-partnership.html

Burns, W. J. (January 9, 2014). International space exploration forum. http://iipdigital.usembassy. gov/st/english/testtrans/2014/01/20140109290196.html. 15 Jan 2020.

Campbell, C. From satellites to the Moon and Mars, China is quickly becoming a super space-power, *Time Magazine*. July 17, 2019. https://time.com/5623537/china-space/

Crane, L. (April 11, 2019). *Israel's Beresheet lunar lander has crashed on the moon*, Space.com. https://www.newscientist.com/article/2199497-israels-beresheet-lunar-lander-has-crashed-on-the-moon/

Frost, J. Japanese Lunar Lander Company on schedule for 2021 first mission, *Space News*. October 24, 2019, mission. https://spacenews.com/japanese-lunar-lander-company-is-pace-on-schedule-for-2021-first-mission/

Howell, E. Space.com. October 25, 2019. NASA's artemis moon program attracts more nations as potential partners, agency says. https://www.space.com/nasa-artemis-moon-program-international-partnerships.html

Indian Mars Orbiter Mission. https://www.space.com/topics/india-mars-orbiter-mission. Last accessed 20 Jan 2020.

Kaguya (Selene), Japanese Aerospace Exploration Agency (JAXA). http://www.kaguya.jaxa.jp/index_e.htm. Last accessed 20 Jan 2020.

Knapton, S. John F Kennedy wanted Moon Mission to be Joint Venture with Soviet Union, *Telegraph*. July 13, 2019. https://www.telegraph.co.uk/science/2019/07/13/john-f-kennedy-wanted-moon-mission-joint-venture-soviet-union/

Kubota, T, Kunii, Y., & Kuroda, Y. *Rover missions and technology for lunar or planetary surface exploration, IEEE*. http://ewh.ieee.org/conf/icra/2008/workshops/PlanetaryRovers/05Kubota/Rover-WS_kubota.pdf#search='SELENEX%20%E3%82%AA%E3%83%97%E3%82%B7%E3%83%A7%E3%83%B3. Last accessed 20 Jan 2020.

Laurini, K. (July 28, 2015). Personal interview. International Space University.

Mahoney, E. (February 2, 2018). International Space Exploration Coordination Group. https://www.nasa.gov/exploration/about/isecg. 2 Feb 2020.

Moskowitz, C. Future Space Exploration Hinges on International Cooperation, Astronauts Say. Space.com. April 29, 2010. http://www.space.com/8297-future-space-exploration-hinges-international-cooperation-astronauts.html. 2 Feb 2020.

Oster, M. Israel space agency gives $5.6 million to help launch second shot at the Moon, *Israel News*. May 8, 2019. https://www.haaretz.com/israel-news/israel-space-agency-gives-5-6-million-to-help-launch-second-shot-at-moon-1.7214622

Pultarova, T. Woerner: Cooperation should reign as spacefaring nations clean up Earth orbit and venture beyond ISS, *Space News*. October 24, 2017. https://spacenews.com/woerner-cooperation-should-reign-as-spacefaring-nations-clean-up-earth-orbit-and-venture-beyond-iss/

Rehm, J. Hope Mars mission: Launching the Arab world into the space race, Space.com. October 21, 2019. https://www.space.com/hope-emirates-mars-mission.html

Results from Chandrayaan-1 Mission. https://www.vssc.gov.in/VSSC/index.php/results-from-chandrayaan-1-mission. Last accessed 20 Jan 2020.

Schmitt, H. *Apollo 17 Astronaut, and one of the last men to step on the surface of the Moon and former New Mexico Senator*. Personal interview. July 28, 2015.

Wall, M. *China eyes robotic outpost at the Moon's south pole in late 2020*, Space.com. July 18, 2019. https://www.space.com/china-moon-south-pole-research-station-2020s.html

Wall, M. India is officially going back to the Moon with Chandrayaan-3 lunar lander, *Space News*. Jan 2, 2020. https://www.space.com/india-confirms-moon-landing-mission-chandrayaan-3.html

Chapter 11
An American Perspective

In order to maintain the United States' status as a world power, America must continue to be the leader in technology, science, and space. In spite of seemingly urgent priorities, daily pressures, and Earthly challenges that face the American people, the United States should remain committed to space exploration. Space activities inspire children to pursue academic excellence, and space-based technologies fuel invention and innovation, and provide many other tangible benefits. These benefits include improved health and medical knowledge, improved cyber-security and safety systems, clean energy, and several other spin-offs which contribute to an overall improved quality of life in society. It would be interesting to see what our life on Earth would be like had the Americans and other nations not pursued a trip to the Moon in the 1960s (Fig. 11.1).

From July 20, 1969, to December 11, 1972, astronauts walked, drove, researched, and even golfed on the surface of the Moon. But when President Nixon canceled the Apollo program in 1972, any prospect of a return mission to the lunar surface became quite bleak. Now, nearly 50 years after leaving that last footprint on the Moon, the US space program finds itself defining strategies and specific timelines. Hopefully, the new policies set forth by the Trump administration will be long-lasting and continue on into future political generations. The lasting imprint of the Nixon space doctrine may be living out its last days (Logsdon 2015).

President George W. Bush's Constellation Program promised astronauts a return to the Moon by 2020. But when President Obama was elected, this vision was terminated following the unveiling of his space policy. Between 2010 and 2017, NASA's human space exploration program remained a vague and aimless plan that provided little or no promise of defined timelines and destinations. Meanwhile, other space-faring nations marched boldly forward as NASA remained spinning in circles, chasing its tail.

On January 20, 2016, when NASA officials admitted that the Space Launch System (SLS), the agency's next big rocket, was a vehicle without a mission plan, NASA acknowledged what was essentially an "empty flight manifest" for the SLS

Fig. 11.1 NASA's Space
Launch System.
(Photo courtesy of NASA)

at NASA's Kennedy Space Center (KSC) during a meeting to discuss the uncertainties facing the SLS. The first scheduled test flight with humans aboard had already been delayed once, and the schedule for future SLS tests was tentative. There was no definitive launch schedule for the rocket after its first unmanned test flight (Grush January 12, 2016).

However, under NASA's current strategy, urgency has become the main focus. When Boeing announced that there would yet be another delay of the first SLS flight manifest, Exploration Mission 1 (now renamed Artemis 1), the President issued Space Policy Directive 1 which called for astronauts to return to the Moon by 2024. This policy challenged both NASA and Boeing to get back on course. The White House was not about to agree with these continual delays and slips in schedules. Vice President Mike Pence immediately flew to NASA's Marshall Spaceflight Center in Huntsville, Alabama. He warned that "NASA is not committed to any one contractor…and if the current contractor cannot meet this objective, then we will find ones that can."

NASA has even considered skipping some of the safety test runs, albeit at the pushback of the Aerospace Safety Advisory Panel (ASAP). This safety organization appointed by Congress considers these tests and milestones necessary for the safety of its crew.

Patricia Sanders, Chair of ASAP, said:

There is no other test approach that will gather the critical full-scale integrated propulsion system operational data required to ensure safe operation. Shorter duration engine firing at the launch pad will not achieve an understanding of the operational margins and could result in severe consequences conducted in a much less controlled environment than Stennis (NASA's rocket testing center is Southern Mississippi) if the margins are exceeded. I cannot emphasize more strongly that we advise NASA to retain this test in the program. (Smith April 25, 2019)

Although these tests can be time-consuming, the proven results will reduce risks and ward off possible life-threatening situations during the actual launches with human crews aboard.

As for now, there remains an unspecified date for the launch of Artemis 1. The SLS rocket and the Orion crew capsule will eventually launch from Kennedy Space Center at the historic Launch Complex 39B. Orion will remain in space for about 3 weeks and will also include several days in lunar orbit in preparation for a human Moon landing. This flight test will be followed by Artemis 2, a crewed mission lasting approximately 10 days. NASA plans to send four astronauts for a lunar flyby and then return safely to Earth. This mission will be the first time humans have traveled beyond low Earth orbit in nearly 50 years (1972).

The good news is that NASA doesn't need to develop much new technology to return to the lunar surface, nor do we need hundreds of billions of dollars. According to John Connolly, it would cost approximately $3–4 billion for a return trip to the Moon. In addition, only a few technologies would need to be developed for longer stays. NASA and its partners need to create life-support systems, learn to extract local available resources, and provide living quarters that shield humans from radiation. As we test these capabilities, it will prepare NASA and its commercial and international partners to move on to Mars mission while taking advantage of the fact that the Moon is a close return journey of only 3 days.

David Kendall, Former Director General of the Canadian Space Agency, and Chairman of the United Nations Committee on the Peaceful Uses of Outer Space (UN COPOUS), commented on NASA's return to the Moon in an interview:

People need inspiration and I believe that space exploration beyond LEO, especially where humans are directly involved, provides this lift. Personally, I am of the opinion that we need to return to the Moon with humans in order to test out the technology and required infrastructure necessary to put humans on Mars. Let us not fool ourselves; space exploration is difficult, but human space exploration, especially beyond LEO is really hard. There is a very important and somewhat impassioned debate, at least in the US, as to how much emphasis should we be putting on a human lunar presence when the ultimate near-term goal is for humans to venture to the red planet. This debate relates to many elements - technology, infrastructure, risk, policy, budgets and politics. States with the wherewithal to send humans beyond LEO – United States, Russia and China – are caught in the dilemma of wanting to be the first, while knowing that a false step could set back their program by many years. Ironically, populations are, for the most part, tolerant of public money being spent on such a high-risk program, but are intolerant of the loss of human life as a consequence. This fact leads to the conclusion that a careful, step-by-step approach is required. However, if it is too conservative, the public will lose interest endangering the large budgets demanded. Personally, my hope is that we can find common ground and, again, use space exploration as a facilitator to dial-down political tensions between the major powers. The remarkable international partnership fostered by the International Space Station is a shining example of what is possible. However, China and India need to be invited as full partners into the next iteration. Will our current political leaders provide this inspiration for the future? (Kendall January 30, 2020)

The Moon and Mars are both very interesting and important destinations in the human exploration effort. It is important that the United States builds on the capabilities and expertise it has today and then incrementally evolves them to meet the challenges that continue. The best approach would be to go to the Moon first while solidifying the partnerships that will help us travel to Mars. The Moon is on the paramount path to Mars, and we can go there with international partners if we can demonstrate the technical abilities together as a unified team. There will be substantial challenges and many problems to solve, and NASA cannot do it alone. But I do not believe NASA wants any other country to lead. Every space agency will need to contribute their capabilities to the missions.

The reality is that each step will not be comparable to Apollo. But with each step we will build consensus that human exploration is worth doing. Space is a strategic domain, and it is worth the sacrifice for the US government and others to invest in space exploration. We inspire, we educate, we drive innovation, we create new knowledge, and we better our life on Earth.

As we return to the Moon, NASA human exploration missions must be independent of the start-stop cycle of the US government's political chess games. Over and over again, we have witnessed the start of a bold American space program, only to find Republican presidents canceling Democrat initiatives and Democrat presidents canceling Republican initiatives. Unfortunately, this pattern gets nothing of significance accomplished. Recently, Congressman Brian Babin said:

> There are thousands of men and women in this country whose days are impacted by the decisions we make in this building. It is easy for people confined to the Beltway Bubble (Washington DC lawmakers) to forget that our pride as Americans comes from the hard work and determination to make this world better. The men and women of NASA working on our human exploration programs are not pawns to be moved around a chess board in the latest game of chicken that the Administration chooses to play with Congress. We must ensure NASA's work focuses on the will of the people, not the political whims of whatever President is in office at the time.

> NASA's human exploration program has been through a tumultuous seven years from 2010 through 2016. We must ensure that there is a constancy of purpose in our planning and a surefooted roadmap in place for the future. Human exploration has a long and storied history of being non-partisan. It is not a Republican or Democrat issue. It is an American issue. We need to get the politics out of this important program. (Babin February 3, 2016)

It is my recommendation that NASA should lead an international partnership similar to the plan that was executed with the International Space Station. If the United States sits back and engages in political competition, China and other spacefaring nations will forge ahead. I am convinced that now is the time for the United States to courageously step up and take charge of a global partnership, with a renewed passion for human space exploration. If America does not, NASA might have to step back and follow someone else.

Findings and Recommendations

NASA's space policy is no longer confused and disorderly, but is well equipped with many visionary goals, which are well-defined strategic plans for future manned missions. The journey to Mars is now an achievable goal in the future, with the critical path of going to the Moon first to prove the necessary technologies. I am a lifelong fan of NASA and human spaceflight and have been so since I was a young girl who watched Neil Armstrong and Buzz Aldrin step foot on the Moon in July of 1969. I would like nothing more than for NASA to go to Mars, and beyond, but what is necessary first is a step-by-step logical schedule of short-term realistic goals. Each milestone must be interesting to the public, even while NASA incrementally builds its space-faring capabilities. This type of plan is sustainable and could be fiscally mapped for Congress. The United States must create a logical, cumulative, affordable plan that leads us to our ultimate destinations. Realizable, short-term space goals will build long-term and enduring space policy and long-term credibility with American taxpayers, Congress, and our international partners.

Eventually, there will be a government that will send astronauts back to the Moon. Will it be China? Russia? The European Space Agency? A joint government-commercial enterprise? I believe NASA, the White House, and Congress will fully realize that this is the progressive, sustainable step to providing the rehearsal needed for successful ventures into deep space. What is truly necessary is bold vision and strong leadership from the White House, with cooperation from Congress, to spur the United States toward adventurous and sustainable achievements in manned space exploration.

For several years there has been continual talk among NASA officials about a journey to Mars, but in the past, there were no real timelines or clear-cut programs in place. Mars is definitely the goal on the horizon, but realistically the initial goal is to acquire the technology and knowledge necessary to live in deep space for long periods of time. The Moon represents a vitally important and progressive step to test our abilities, which will lead us all to a successful trip to the red planet.

On March 15, 2016, Congressman John Culberson (R-TX), Chairman of the Committee on Appropriations Subcommittee on Commerce, Justice, Science, and Related Agencies, stated his views on the FY2017 NASA budget:

> NASA just accepted a new group of applications for astronauts with over 18,300 applications for 14 astronaut spots. That is an indication of the level of support the country has for the space program. Every time there is a major mission launch the NASA website becomes one of the busiest, and OMB (Office of Management and Budget)] refuses to give you the support you deserve. This committee will make sure you get the budget you need. It will be a tough budget year but we will be behind you every step of the way. It is frustrating when we love NASA and want you to stay the course. (Culberson March 15, 2016)

US space policy should contain several attributes in order to maintain a consistent tempo from administration to administration. The missions should demonstrate to taxpaying citizens, Congress, and industry that there is continual progression in human space exploration. This will produce confidence in NASA's long-term strategy and in the leaders of our country to perform what was promised. NASA should continue to choose destinations that are inspirational and reflect significant scientific, economic, and geopolitical advantages. The pathway forward should also reflect responsible progression, both fiscally and technologically.

On February 25, 2016, Culberson gave his assessment of the current situation at NASA:

> We need to make NASA less political, more professional, and give them the ability to see far into the future with knowledge and confidence that the Congress will be there behind them. Over the last twenty years NASA has spent more than twenty billion dollars on cancelled development programs. No company, no entity, no agency of the federal government can function in this environment. (Smith February 29, 2016)

Also commenting at the same hearing was Eileen Collins, first female Commander of the Space Shuttle:

> Program cancellations made by bureaucracies behind closed doors, without input by the people, are divisive, damaging, cowardly, and many times more expensive in the long run. A continuity of purpose over many years and political administrations will avoid surprises that set us back years. (Collins February 25, 2016)

Human spaceflight and space itself are still tremendous sources of inspiration for Americans. It is essential that future presidential administrations tap into that pride and maintain the United States' leadership in space. It is a meaningful part of our history, and the public is becoming more excited about making this happen again. But it is going to require committed leadership in the White House, with genuine dedication and unwavering persistence, coupled with follow-through. The next president will have to consider the geopolitical and fiscal situations, set up a winning formula, and then passionately communicate this to the American people, which will, in turn, be passed along to the NASA organization and its partners.

The Moon and Mars are interesting and important destinations in the human exploration effort. NASA needs to build on the capabilities and expertise that we have today and evolve them incrementally to meet the challenges that face us. The best approach would be to go to the Moon first and then solidify partnerships that will help us get to Mars. The Moon is on the critical path to Mars, and we can go there with international partners if we can demonstrate the technical abilities together as a unified team. There will be significant challenges and many problems to solve and NASA can't do it alone. Every space agency will need to contribute their capabilities.

At present, the majority of international partners do not want to go to Mars; they want to go to the Moon. There is a strong relationship between NASA and its international space partners. While it is true that the United States has set foot on the Moon, other nations have not yet done so. It seems likely that the rationale for their decisions comes down to money and affordability. With most national space budgets under severe restraint through budget cuts, the next logical step (given the realities of the geopolitical landscape) strongly suggests a return to the Moon as most nations' first priority.

If we wish to advance the cause of global peace and prosperity, it is vital that we make shared space exploration an international collaboration. In doing so, we as a species will accelerate our progress here on Earth while we continue to strive to unlock the mysteries of our vast universe.

References

Babin, B. (February 3, 2016). *Charting a course: Expert perspectives on NASA's human exploration*. Hearing of the House Committee on Science, Space and Technology Subcommittee on Space.

Collins, E. (February 25, 2016). Hearing of the House, Science, Space and Technology Committee.

Culberson, J. (March 15, 2016). *Budget hearing, NASA FY 2017*. Committee on Appropriations Subcommittee on Commerce, Justice, Science, and Related Agencies.

Grush, L. NASA officials admit space launch system is a rocket without a plan. *The Verge.* January 12, 2016. http://www.theverge.com/2016/1/12/1075811/nasa-ksc-meetings-sls-rocket-uncertain-launch-dates. 12 Sept 2019.

Kendall, D. (January 30, 2020). Chair United Nations COPUOS 2016–2017, Former Director General of the Canadian Space Agency. Personal interview, Director International Space University.

Logsdon, J. M. (2015). *After Apollo*. London: Palgrave Macmillan.

Smith, M. S. Witnesses support goal of NASA restructuring legislation, but not specifics. *Space Policy Online.* February 29, 2016.

Smith, M. S. Safety panel emphatically urges NASA not to skip SLS green run test. *Space Policy Online.* April 25, 2019.

Chapter 12
The Way Forward

NASA's next logical step in exploring beyond low Earth orbit is manned missions to the Moon. It is a natural satellite and possibly a future research laboratory that is close and relatively accessible. Its close proximity makes it a rational proving ground for future manned mission to Mars and beyond. Equipment, systems, and hardware could be proven and tested to check their reliability in a radiation-rich environment. The Moon's reduced gravity is similar enough to Mars to test human performance and yield the knowledge needed to make sure that our astronauts will be safe. Mars simulations could easily be performed in long-duration visits realistically assessing adaptability of the crew, both physiologically and psychologically. Surface hardware including habitation structures, life-support systems, power generators, mobile vehicles, and in situ resources could be developed and matured. Everything that is accomplished on the Moon should reflect a cumulative, layered, stair step process toward a successful crewed mission to Mars (Fig. 12.1).

Canadian astronaut Chris Hadfield has flown three space missions and served as Commander of the ISS. He was inspired to become an astronaut when he watched the Apollo 11 Moon landing on television from a small farm in southern Ontario, Canada. Hadfield is anxious for people to venture out into deep space, but he believes the next logical step is returning to the Moon and building lunar colonies:

> We will be on the International Space Station for another ten years or so, and where is the next obvious place we'll go? The Moon. It's only three days away. . . . The Moon is for ideal for testing . . . . we need to be able to get everything right and not kill everybody. (Matt Burgess, January 22, 2016)

When NASA went to the Moon from 1969 to 1972, it was for footprints and flags. The few astronauts who went spent very little time there, and that was nearly 50 years ago. The "New Moon" vision will be to stay for long periods of time and learn to actually live there. During these lengthy missions, multitudes of technologies will be tested that will eventually be needed for life in deep space and travel

A. Reneau, *Moon First and Mars Second*, SpringerBriefs in Space Development, https://doi.org/10.1007/978-3-030-54230-6_12

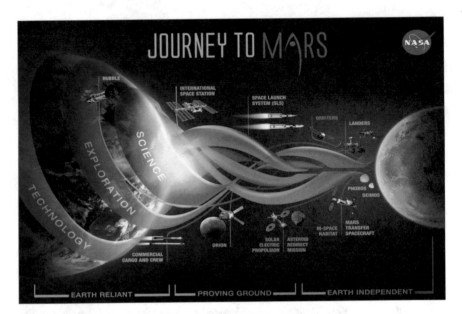

Fig. 12.1 NASA's journey to Mars. (Graphic courtesy of NASA)

throughout the solar system. As the clock ticks on the termination of the ISS, it is time to plan the return to the Moon. There is no doubt that this will be NASA's next logical leap beyond low Earth orbit.

Geopolitical competition always fuels rapid advances in technology; unfortunately, most of today's major technological advances have been triggered by threats of some sort. Ultimately, it could well be that the threat of competition for dominance in deep space could drive the next space missions. When citizens do not feel safe, suddenly an availability of funds will appear. It is not clear when all of these conditions will intersect. Likewise, it is far from clear from whence the finances, the motivation, and the common vision for missions to the Moon and beyond will ultimately come. One must hope that a sustainable schedule of human spaceflight missions both now and future generations will eventually be achieved, and the author believes that successful human deep-space exploration will reveal the ultimate human destiny.

When Apollo 17 Astronaut Gene Cernan boarded the lunar module for the last time, and left the last footprint on the Moon, on December 13, 1972, he said:

> I'm on the surface; and, as I take man's last step from the surface, back home for some time to come — but we believe not too long into the future — I'd like to just [say] what I believe history will record. That America's challenge of today has forged man's destiny of tomorrow. And, as we leave the Moon at Taurus-Littrow, we leave as we came and, God willing, as we shall return: with peace and hope for all mankind. Godspeed the crew of Apollo 17. (Eugene A. Cernan December 13, 1972)

References

Burgess, M. (January 22, 2016). *Chris Hadfield: Moon colonization is obvious next step.* Wired. Co. UK. http://www.wired.co.uk/news/archive/2016-01/22/chris-hadfield-Moon-Mars-spacex. Last accessed 21 Oct 2019.

Cernan, E. http://www.nmspacemuseum.org/halloffame/detail.php?id=61. Last accessed 18 Feb 2020.

Appendices

Space Policy Directive 1

PRESIDENTIAL MEMORANDA

Presidential Memorandum on Reinvigorating America's Human Space Exploration Program

INFRASTRUCTURE & TECHNOLOGY

Issued on: December 11, 2017

ALL NEWS

MEMORANDUM FOR THE VICE PRESIDENT
THE SECRETARY OF STATE
THE SECRETARY OF DEFENSE
THE SECRETARY OF COMMERCE
THE SECRETARY OF TRANSPORTATION
THE SECRETARY OF HOMELAND SECURITY
THE DIRECTOR OF NATIONAL INTELLIGENCE
THE DIRECTOR OF THE OFFICE OF MANAGEMENT AND BUDGET
THE ASSISTANT TO THE PRESIDENT FOR NATIONAL SECURITY AFFAIRS
THE ADMINISTRATOR OF THE NATIONAL AERONAUTICS AND SPACE
    ADMINISTRATION
THE DIRECTOR OF THE OFFICE OF SCIENCE AND TECHNOLOGY POLICY
THE ASSISTANT TO THE PRESIDENT FOR HOMELAND SECURITY AND
    COUNTERTERRORISM
THE CHAIRMAN OF THE JOINT CHIEFS OF STAFF

SUBJECT: Reinvigorating America's Human Space Exploration Program
Section 1. Amendment to Presidential Policy Directive-4. [President Barack Obama]

© The Editor(s) (if applicable) and The Author(s), under exclusive license to
Springer Nature Switzerland AG 2021
A. Reneau, *Moon First and Mars Second*, SpringerBriefs in Space
Development, https://doi.org/10.1007/978-3-030-54230-6

Presidential Policy Directive-4 of June 28, 2010 (National Space Policy), is amended as follows:

The paragraph beginning "Set far-reaching exploration milestones" is deleted and replaced with the following:

Lead an innovative and sustainable program of exploration with commercial and international partners to enable human expansion across the solar system and to bring back to Earth new knowledge and opportunities. Beginning with missions beyond low-Earth orbit, the United States will lead the return of humans to the Moon for long-term exploration and utilization, followed by human missions to Mars and other destinations;.

Sec. 2. General Provisions.

(a) Nothing in this memorandum shall be construed to impair or otherwise affect:

    (i) the authority granted by law to an executive department or agency, or the head thereof; or

    (ii) the functions of the Director of the Office of Management and Budget relating to budgetary, administrative, or legislative proposals.

(b) This memorandum shall be implemented consistent with applicable law and subject to the availability of appropriations.

(c) This memorandum is not intended to, and does not, create any right or benefit, substantive or procedural, enforceable at law or in equity by any party against the United States, its departments, agencies, or entities, its officers, employees, or agents, or any other person.

(d) This memorandum shall be published in the Federal Register.

DONALD J. TRUMP

Artemis Overview by NASA

Artemis

July 25, 2019

What Is Artemis?

Artist's concept of the Space Launch System rocket and Orion capsule prepared for launch
  Credits: NASA

NASA is committed to landing American astronauts, including the first woman and the next man, on the Moon by 2024. Through the agency's Artemis lunar exploration program, we will use innovative new technologies and systems to explore more of the Moon than ever before. We will collaborate with our commercial and international partners to establish sustainable missions by 2028. And then we will use what we learn on and around the Moon to take the next giant leap—sending astronauts to Mars.

Why Go to the Moon?

With the Artemis program we will:

- Demonstrate new technologies, capabilities, and business approaches needed for future exploration including Mars
- Establish American leadership and a strategic presence on the Moon while expanding our US global economic impact
- Broaden our commercial and international partnerships
- Inspire a new generation and encourage careers in STEM

How Do We Get There?

NASA's powerful new rocket, the Space Launch System (SLS), will send astronauts aboard the Orion spacecraft nearly a quarter million miles from Earth to lunar orbit. Astronauts will dock Orion at the Gateway and transfer to a human landing system for expeditions to the surface of the Moon. They will return to the orbital outpost to board Orion again before returning safely to Earth.

When Will We Get There?

Ahead of the human return, we will send a suite of science instruments and technology demonstrations to the lunar surface through commercial Moon deliveries beginning in 2021.

The agency will fly two missions around the Moon to test its deep-space exploration systems. NASA is working toward launching Artemis I, an uncrewed flight to test the SLS and Orion spacecraft together, followed by the Artemis II mission, the first SLS and Orion test flight with crew. NASA will land astronauts on the Moon by 2024 on the Artemis III mission and about once a year thereafter.

Artist's concept of the Phase 1 Gateway in lunar orbit with Orion and the human landing system docked to the orbital outpost
Credits: NASA

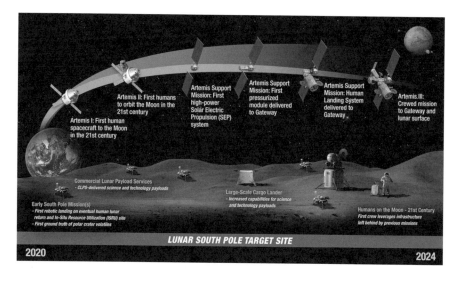

Download the Artemis Fact Sheet

What Will We Do There?

While Mars remains our horizon goal, we have set our sights first on exploring the entire surface of the Moon with human and robotic explorers. We will send astronauts to new locations, starting with the lunar South Pole. At the Moon, we will:

- Find and use water and other critical resources needed for long-term exploration
- Investigate the Moon's mysteries and learn more about our home planet and the universe
- Learn how to live and operate on the surface of another celestial body where astronauts are just 3 days from home

Prove the technologies we need before sending astronauts on missions to Mars, which can take up to 3 years round trip

Index

Printed in the United States
By Bookmasters